Chemical Elements

2nd Edition

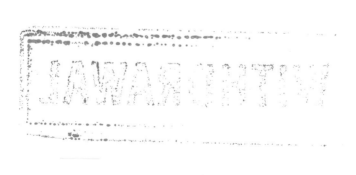

Chemical Elements

2nd Edition VOLUME 2: G-O

David E. Newton
Kathleen J. Edgar, Editor

U·X·L
A part of Gale, Cengage Learning

GALE
CENGAGE Learning

Detroit • New York • San Francisco • New Haven, Conn • Waterville, Maine • London

GALE
CENGAGE Learning

Chemical Elements, 2nd Edition

David E. Newton

Project Editor: Kathleen J. Edgar

Managing Editor: Debra Kirby

Rights Acquisition and Management: Barbara McNeil, Jackie Jones, and Robyn Young

Imaging and Multimedia: John L. Watkins

Composition: Evi Abou-El-Seoud, Mary Beth Trimper

Manufacturing: Wendy Blurton, Dorothy Maki

Product Managers: Julia Furtaw, John Boyless

Product Design: Jennifer Wahi

For product information and technology assistance, contact us at **Gale Customer Support, 1-800-877-4253.**
For permission to use material from this text or product, submit all requests online at **www.cengage.com/permissions.**
Further permissions questions can be e-mailed to **permissionrequest@cengage.com.**

Cover photographs: Beaker image by originalpunkt, used under license from Shutterstock.com. All other photos by Photos.com/Getty Images.

While every effort has been made to ensure the reliability of the information presented in this publication, Gale, a part of Cengage Learning, does not guarantee the accuracy of the data contained herein. Gale accepts no payment for listing; and inclusion in the publication of any organization, agency, institution, publication, service, or individual does not imply endorsement of the editors or publisher. Errors brought to the attention of the publisher and verified to the satisfaction of the publisher will be corrected in future editions.

LIBRARY OF CONGRESS CATALOGING-IN-PUBLICATION DATA

Newton, David E.
  Chemical elements / David E. Newton ; Kathleen J. Edgar, editor. -- 2nd ed.
    p. cm.
  Includes bibliographical references and index.
  ISBN-13: 978-1-4144-7608-7 (set)
  ISBN-13: 978-1-4144-7609-4 (v. 1)
  ISBN-13: 978-1-4144-7610-0 (v. 2)
  ISBN-13: 978-1-4144-7611-7 (v. 3)
  [etc.]
  1. Chemical elements. I. Edgar, Kathleen J. II. Title.
  QD466.N464 2010
  546--dc22   2009050931

Gale
27500 Drake Rd.
Farmington Hills, MI, 48331-3535

ISBN-13: 978-1-4144-7608-7 (set)   ISBN-10: 1-4144-7608-6 (set)
ISBN-13: 978-1-4144-7609-4 (vol. 1)   ISBN-10: 1-4144-7609-4 (vol. 1)
ISBN-13: 978-1-4144-7610-0 (vol. 2)   ISBN-10: 1-4144-7610-8 (vol. 2)
ISBN-13: 978-1-4144-7611-7 (vol. 3)   ISBN-10: 1-4144-7611-6 (vol. 3)

This title is also available as an e-book.
ISBN-13: 978-1-4144-7612-4 (set)   ISBN-10: 1-4144-7612-4 (set)
Contact your Gale sales representative for ordering information.

Printed by China Translation & Printing Services Limited, Guangdong Province, China. 2nd printing. 07/2011
2 3 4 5 6 7 14 13 12 11

Main–Group Elements

1
I A

Atomic number | 86 | [222] | Atomic weight
Symbol | **Rn**
Name | radon

			1	1.00794											

Period

Period	1 I A	2 II A	3 III B	4 IV B	5 V B	6 VI B	7 VII B	8	9 VIII B
1	**1** 1.00794 **H** hydrogen								
2	**3** [6.941] **Li** lithium	**4** 9.012182 **Be** beryllium							
3	**11** 22.98976928 **Na** sodium	**12** 24.3050 **Mg** magnesium							
4	**19** 39.0983 **K** potassium	**20** 40.078 **Ca** calcium	**21** 44.955912 **Sc** scandium	**22** 47.867 **Ti** titanium	**23** 50.9415 **V** vanadium	**24** 51.9961 **Cr** chromium	**25** 54.938045 **Mn** manganese	**26** 55.845 **Fe** iron	**27** 58.933195 **Co** cobalt
5	**37** 85.4678 **Rb** rubidium	**38** 87.62 **Sr** strontium	**39** 88.90585 **Y** yttrium	**40** 91.224 **Zr** zirconium	**41** 92.90638 **Nb** niobium	**42** 95.96 **Mo** molybdenum	**43** [98] **Tc** technetium	**44** 101.07 **Ru** ruthenium	**45** 102.90550 **Rh** rhodium
6	**55** 132.9054519 **Cs** cesium	**56** 137.327 **Ba** barium	**57–71** * lanthanoids	**72** 178.49 **Hf** hafnium	**73** 180.94788 **Ta** tantalum	**74** 183.84 **W** tungsten	**75** 186.207 **Re** rhenium	**76** 190.23 **Os** osmium	**77** 192.217 **Ir** iridium
7	**87** [223] **Fr** francium	**88** [226] **Ra** radium	**89–103** † actinoids	**104** [261] **Rf** rutherfordium	**105** [262] **Db** dubnium	**106** [266] **Sg** seaborgium	**107** [264] **Bh** bohrium	**108** [277] **Hs** hassium	**109** [268] **Mt** meitnerium

Transition Metals

*Lanthanoids

57 138.90547 **La** lanthanum	**58** 140.116 **Ce** cerium	**59** 140.90765 **Pr** praseodymium	**60** 144.242 **Nd** neodymium	**61** [145] **Pm** promethium	**62** 150.36 **Sm** samarium

†Actinoids

89 [227] **Ac** actinium	**90** 232.03806 **Th** thorium	**91** 231.03588 **Pa** protactinium	**92** 238.02891 **U** uranium	**93** [237] **Np** neptunium	**94** [244] **Pu** plutonium

Color Key

- alkali metals
- alkaline earth metals
- transition metals
- inner-transition metals
- metalloids
- other metals
- other non-metals
- halogens
- noble gases
- under review
- not yet discovered

Main-Group Elements

			18 VIII A
			2 4.002602 **He** helium

13 III A	14 IV A	15 V A	16 VI A	17 VII A	
5 10.811 **B** boron	6 12.0107 **C** carbon	7 14.0067 **N** nitrogen	8 15.9994 **O** oxygen	9 18.9984032 **F** fluorine	10 20.1797 **Ne** neon

10	11 I B	12 II B	13 26.9815386 **Al** aluminum	14 28.0855 **Si** silicon	15 30.973762 **P** phosphorus	16 32.065 **S** sulfur	17 35.453 **Cl** chlorine	18 39.948 **Ar** argon
28 58.6934 **Ni** nickel	29 63.546 **Cu** copper	30 65.38 **Zn** zinc	31 69.723 **Ga** gallium	32 72.64 **Ge** germanium	33 74.92160 **As** arsenic	34 78.96 **Se** selenium	35 79.904 **Br** bromine	36 83.798 **Kr** krypton
46 106.42 **Pd** palladium	47 107.8682 **Ag** silver	48 112.411 **Cd** cadmium	49 114.818 **In** indium	50 118.710 **Sn** tin	51 121.760 **Sb** antimony	52 127.60 **Te** tellurium	53 126.90447 **I** iodine	54 131.293 **Xe** xenon
78 195.084 **Pt** platinum	79 196.966569 **Au** gold	80 200.59 **Hg** mercury	81 204.3833 **Tl** thallium	82 207.2 **Pb** lead	83 208.98040 **Bi** bismuth	84 [209] **Po** polonium	85 [210] **At** astatine	86 [222] **Rn** radon
110 [271] **Ds** darmstadtium	111 [272] **Rg** roentgenium	112 [285] **Cn** copernicium	113 [284] **Uut** ununtrium	114 [289] **Uuq** ununquadium	115 [288] **Uup** ununpentium	116 [293] **Uuh** ununhexium	117 **Uus** ununseptium	118 [294] **Uuo** ununoctium

Inner-Transition Metals

63 151.964 **Eu** europium	64 157.25 **Gd** gadolinium	65 158.92535 **Tb** terbium	66 162.500 **Dy** dysprosium	67 164.93032 **Ho** holmium	68 167.259 **Er** erbium	69 168.93421 **Tm** thulium	70 173.054 **Yb** ytterbium	71 174.9668 **Lu** lutetium
95 [243] **Am** americium	96 [247] **Cm** curium	97 [247] **Bk** berkelium	98 [251] **Cf** californium	99 [252] **Es** einsteinium	100 [257] **Fm** fermium	101 [258] **Md** mendelevium	102 [259] **No** nobelium	103 [262] **Lr** lawrencium

Contents

Contents

Contents: Elements by Atomic Number

Chemical Elements, 2nd Edition

Contents: Elements by Family Group

Reader's Guide

Many young people like to play with Lego blocks, erector sets, and similar building games, including virtual games found on the Internet. It's fun to see how many different ways a few simple shapes can be put together.

The same can be said of chemistry. The world is filled with an untold number of different objects, ranging from crystals and snowflakes to plant and animal cells to plastics and medicines. Yet all of those objects are made from various combinations of only about 100 basic materials: the chemical elements.

Scientists have been intrigued about the idea of an "element" for more than 2,000 years. The early Greeks developed complicated schemes that explained everything in nature using only a few basic materials, such as earth, air, fire, and water. The Greeks were wrong in terms of the materials they believed to be "elemental." But they were on the right track in developing the concept that such materials did exist.

By the 1600s, chemists were just beginning to develop a modern definition of an element. An element, they said, was any object that cannot be reduced to some simpler form of matter. Over the next 300 years, research showed that about 100 such materials did exist. These materials range from such well-known elements as oxygen, hydrogen, iron, gold, and silver to substances that are not at all well known, elements such as neodymium, terbium, rhenium, seaborgium, darmstadtium, and copernicium.

By the mid-1800s, the search for new chemical elements had created a new problem. About 50 elements were known at the time. But no one yet knew how these different elements related to each other, if they did at all. Then, in one of the great coincidences in chemical history, that question was answered independently by two scientists at almost the same time, German chemist Lothar Meyer (1830–1895) and Russian chemist Dmitri Mendeleev (1834–1907). (Meyer, however, did not publish his research until 1870, nor did he predict the existence of undiscovered elements as Mendeleev did.)

Meyer and Mendeleev discovered that the elements could be grouped together to make them easier to study. The grouping occurred naturally when the elements were laid out in order, according to their atomic weight. Atomic weight is a quantity indicating atomic mass that tells how much matter there is in an element or how dense it is. The product of Meyer and Mendeleev's research is one of the most famous visual aids in all of science, the periodic table. Nearly every classroom has a copy of this table. (A copy is available in this book just before the table of contents.) It lists all of the known chemical elements, arranged in rows and columns. The elements that lie within a single column or a single row all have characteristics that relate to each other. Chemists and students of chemistry use the periodic table to better understand individual elements and the way the elements are similar to and different from each other.

About *Chemical Elements, 2nd Edition*

Chemical Elements, 2nd Edition is designed as an introduction to the chemical elements. This new edition, presented in full color, updates the earlier three-volume set, providing new information about the elements as well as many new photographs.

Students will find *Chemical Elements* useful in a number of ways. First, it is a valuable source of fundamental information for research reports, science fair projects, classroom demonstrations, and other activities. Second, it provides more detail about elements and compounds that are only mentioned in other science textbooks or classrooms. Third, it is an interesting source of information about the building blocks of nature for those who simply want to know more about the elements.

Elements with atomic numbers 1 through 100 are presented in separate entries. The transfermium elements (elements 101 through 112) are covered in one entry, which also discusses six additional elements (113, 114, 115, 116, 118, and 122) that have yet to be confirmed by the International Union of Pure and Applied Chemistry (IUPAC).

Entry Format

Chemical Elements is arranged alphabetically by element name. All entries open with an overview section designed to introduce students to the basics of the element. Each entry then contains specific information in the following categories:

- Discovery and Naming
- Physical Properties
- Chemical Properties
- Occurrence in Nature
- Isotopes
- Extraction
- Uses
- Compounds
- Health Effects

In addition, the first page of each entry features a Key Facts section in the margin. Here, the element's chemical symbol, atomic number, atomic mass, family, and pronunciation are shown. A diagram of an atom of the element is also shown at the top of the page. In the diagram, the atom's electrons are arranged in various "energy levels" outside the nucleus. Inside the nucleus, the number of protons and neutrons is indicated.

Entries are easy to read and written in a straightforward style. Difficult words are defined within the text. Each entry also includes a "Words to Know" sidebar that defines technical words and scientific terms. This enables students to learn vocabulary appropriate to chemistry without having to consult other sources for definitions.

Special Features

Chemical Elements, 2nd Edition includes a number of special features that help make the connection between the elements, minerals, people who discovered and worked with them, and common uses of the elements. These features include:

- Nearly 300 photographs and illustrations, most in color, showcasing various elements, the products in which they are used, and the places in which they are found. Such imagery brings the elements to life. Black-and-white images of historical figures are also included.

- Sidebars discussing fascinating supplemental information about scientists, theories, uses of elements, and more.
- Extensive cross-references. Other elements mentioned within an entry appear in bold type upon first mention, serving as a helpful reminder that separate entries are written about these other elements.
- A periodic table (positioned right before the table of contents) that includes the following for each element: name, symbol, atomic number, and atomic mass. A color key also informs students about the various groupings of the elements.
- Three tables of contents—alphabetically by element name, by atomic number, and by family group. This presentation provides varied access to the elements.
- A timeline in each volume. This section details the chronology of the discovery of the elements.
- A cumulative Words to Know section in each volume. This section collects all the glossary terms used in the individual entries.
- A Where to Learn More section at the back of each volume. This bibliography provides information on books, periodicals, Web sites, and organizations that may be useful to those interested in learning more about the chemical elements.
- A comprehensive index that quickly points readers to the elements, minerals, and people mentioned in *Chemical Elements, 2nd Edition*.

A Note about Isotopes and Atomic Weights

Various authorities list slightly different isotopes for some elements. One reason is that the discovery of a new isotope may not yet have been confirmed by other researchers, so its existence is uncertain. Lists of isotopes may change also because new isotopes are being discovered from time to time. The number of isotopes, with a known half life, listed in this book is based on information available from the Lawrence Berkeley National Laboratory in Berkeley, California. The list is up to date as of December 2009. All stable isotopes exist in nature. Some radioactive isotopes also exist in nature, but the vast majority has been prepared synthetically in the laboratory.

The atomic weight information presented in the entries and periodic table is the latest available (at press time) from the International Union of

Pure and Applied Chemistry (IUPAC), the governing body that officially confirms and names any new elements. The IUPAC lists the atomic weight in brackets when an element does not have any stable nuclides. For example, the atomic weight of actinium is listed as [227]. This represents the mass of its longest-lived isotope.

Special Thanks

The project editor wishes to thank graphic artist Christine O'Bryan for the amazing job she did colorizing the atom diagrams and updating the periodic table. Thanks also to Robyn Young for working her magic in acquiring new images for this edition. Also thanks to Lemma Shomali for lending her content expertise to this set. All your efforts are greatly appreciated.

Comments and Suggestions

We welcome your comments on this work as well as suggestions for future science titles. Please write: Editors, *Chemical Elements, 2nd Edition*, U•X•L, 27500 Drake Rd., Farmington Hills, MI, 48331-3535; call toll-free: 800-347-4253; send fax to 248-699-8066; or send e-mail via http://www.gale.com.

Timeline: The Discovery of Elements

Assigning credit for the discovery of a new element is often a difficult and complicated process. First, many elements were in use well before recorded history. In some cases, these elements were known in the form of their compounds, but not as pure elements. Elements that fall into this category include carbon, copper, gold, iron, lead, mercury, silver, sulfur, tin, and, perhaps, zinc.

In addition, the discovery of an element has seldom been a single, clear-cut event that occurs in such a way that everyone agrees that "X" should receive credit for discovering the element. Instead, the first step in the process of discovery is often the recognition that a new substance has been found-—a new mineral, rock, compound, or other material—that has properties different from anything previously known. This discovery may lead a chemist (or a number of chemists) to suspect the existence of a new element.

The next step may be to isolate the element, either in its pure form or, more commonly, as a compound, such as the oxide or sulfide of the new element. Finally, someone is able to prepare a pure sample of the element, which the world then sees for the first time. An example of this sequence of events can be seen in the elements that make up groups 1 and 2 of the periodic table. Most of those elements were known in one form or another for centuries. But it was not until the early 1800s that Sir Humphry Davy found a method for isolating the pure elements from their oxides.

The process becomes even more complicated when a truly new element is discovered which, sometime later, is found not to be a single element, but a mixture of two or more new elements. The story of the discovery of the rare earth elements is probably the best example of this process.

This sequence of events often takes place over an extended period of time, many years or even decades. For that reason, assigning a specific date to the discovery of an element can also be difficult. Does one choose the date and person when a new compound of the element is discovered, when the pure element itself is prepared, when the discoverer publicly announces his or her discovery, or when official confirmation of the discovery is announced?

For all these reasons, the dates and names listed below must be considered as somewhat ambiguous. For more detailed information about the discovery of each element, the reader should refer to the entry for that element in the main body of this set of books.

About 800 CE Persian natural philosopher Abu Musa Jābir ibn Hayyān al azdi (better known as Geber) is credited with discovering **antimony, arsenic,** and **bismuth**.

About 800 CE Indian metallurgist Rasaratna Samuchaya is perhaps the first person to recognize **zinc** as an element.

1250 German natural philosopher Albertus Magnus is credited as being the first European to discover **arsenic**.

About 1450 The apocryphal Basilius Valentinus (Basil Valentine) is the first European to mention elemental **antimony** and **bismuth**.

1526 Swiss physician Auroleus Phillipus Theostratus Bombastus von Hohenheim (Paracelsus) is acknowledged as the modern discoverer of **zinc**.

1669 German physician Hennig Brand discovers **phosphorus**.

1735 Swedish chemist Georg Brandt discovers **cobalt**.

1735–1748 Spanish military Leader Don Antonio de Ulloa discovers **platinum**.

1751 Swedish mineralogist Axel Fredrik Cronstedt discovers **nickel**.

1755 Scottish physician and chemist Joseph Black recognizes the presence of a new element in magnesia alba, later found to be **magnesium**.

1766 English chemist and physicist Henry Cavendish discovers **hydrogen**.

1771 Swedish chemist Carl Wilhelm Scheele discovers **oxygen**, but does not publish his discovery until 1777.

1772 Scottish physician and chemist Daniel Rutherford discovers **nitrogen**.

1774 Swedish chemist Carl Wilhelm Scheele discovers **chlorine**.

1774 Swedish mineralogist Johann Gottlieb Gahn discovers **manganese**.

1774 English chemist Joseph Priestley discovers **oxygen** and, because he announces his results almost immediately, is often credited as the discoverer of the element.

1781 Swedish chemist Peter Jacob Hjelm discovers **molybdenum**.

1782 Austrian mineralogist Baron Franz Joseph Müller von Reichenstein discovers **tellurium**.

1783 Spanish scientists Don Fausto D'Elhuyard and Don Juan José D'Elhuyard and Swedish chemist Carl Wilhelm Scheele discover **tungsten**.

1787 Scottish military surgeon William Cruikshank and Irish chemist and physicist Adair Crawford independently announce the probable existence of a new element, later found to be **strontium**.

1789 German chemist Martin Klaproth recognizes the presence of **uranium** in pitchblende, but does not isolate the element.

1789 German chemist Martin Klaproth discovers **zirconium**. The element is not isolated until 1824.

1791 English clergyman William Gregor discovers an oxide of **titanium**. German chemist Martin Klaproth makes a similar discovery four years later. The element is not isolated until 1910.

1794 Finnish chemist Johan Gadolin discovers **yttrium**.

1797 French chemist Louis-Nicolas Vauquelin discovers **chromium**.

1798 French chemist Louis-Nicolas Vauquelin discovers **beryllium**.

1801 English chemist Charles Hatchett discovers **niobium**.

1801 Spanish-Mexican metallurgist Andrés Manuel del Río discovers **vanadium**.

1802 Swedish chemist and mineralogist Anders Gustaf Ekeberg discovers **tantalum**.

1803 English chemist and physicist William Hyde Wollaston discovers **palladium**.

1803 Swedish chemists Jöns Jakob Berzelius and Wilhelm Hisinger and German chemist Martin Klaproth discover the black rock of Bastnas, Sweden, which led to the discovery of several elements. Berzelius and Hisinger originally assume the rock is a new element, which they name **cerium**.

1803 English chemist Smithson Tennant discovers **osmium** and **iridium**.

1804 English chemist and physicist William Hyde Wollaston discovers **rhodium**.

1807–1808 English chemist Sir Humphry Davy isolates a number of elements in a pure form for the first time, including **potassium, sodium, magnesium, barium, calcium**, and **strontium**.

1808 French chemists Louis Jacques Thênard and Joseph Louis Gay-Lussac discover **boron**. Davy isolates the element a few days after its discovery has been announced.

1811 French chemist Bernard Courtois discovers **iodine**.

1817 Swedish chemist Johan August Arfwedson discovers **lithium**.

1817 German chemists Friedrich Stromeyer, Karl Samuel Leberecht Hermann, and J. C. H. Roloff independently discover **cadmium**, a name chosen by Stromeyer.

1818 Swedish chemists Jöns Jakob Berzelius and J. G. Gahn discover **selenium**.

1823 Swedish chemist Jöns Jakob Berzelius discovers **silicon**.

1825 Danish chemist and physicist Hans Christian Oersted discovers **aluminum**.

1825 French chemist Antoine-Jérôme Balard and German chemist Leopold Gmelin independently discover **bromine**.

1829 Swedish chemist Jöns Jakob Berzelius discovers **thorium**.

1830 Swedish chemist Nils Gabriel Sefström rediscovers **vanadium**.

1838 Swedish chemist Carl Gustav Mosander discovers that **cerium** contains a new element, which he names **lanthanum**. His lanthanum is later found to consist of four new elements.

1842 Swedish chemist Carl Gustav Mosander discovers that the earth called yttria actually consists of two new elements, **erbium** and **terbium**.

1844 Russian chemist Carl Ernst Claus discovers **ruthenium**.

1860 German chemists Robert Bunsen and Gustav Kirchhoff discover **cesium**.

1861 German chemists Robert Bunsen and Gustav Kirchhoff discover **rubidium**.

1861 British physicist Sir William Crookes discovers **thallium**.

1863 German chemists Ferdinand Reich and Hieronymus Theodor Richter discover **indium**.

1868 Pierre Janssen and Norman Lockyer discover **helium** in the spectrum of the sun.

1875 Paul-Émile Lecoq de Boisbaudran discovers **gallium**.

1878 Swiss chemist Jean-Charles-Galissard de Marignac, Swedish chemist Lars Fredrik Nilson, and French chemist Georges Urbain all receive partial credit for the discovery of **ytterbium**.

1878–1879 Swedish chemist Per Teodor Cleve discovers **holmium** and **thulium**.

1879 Swedish chemist Lars Fredrik Nilson discovers **scandium**.

1880 French chemist Paul-Émile Lecoq de Boisbaudran discovers **samarium**.

1880 French chemist Jean-Charles-Galissard de Marignac discovers **gadolinium**.

1885 Austrian chemist Carl Auer von Welsbach discovers **praseodymium** and **neodymium**.

1885 German chemist Clemens Alexander Winkler discovers **germanium**.

1886 French chemist Henri Moissan discovers **fluorine**.

1886 French chemist Paul-Émile Lecoq de Boisbaudran discovers **dysprosium**.

1894 English chemists Lord Rayleigh and Sir William Ramsay discover **argon**.

1895 English chemist Sir William Ramsay and Swedish chemists Per Teodor Cleve and Nils Abraham Langlet independently discover **helium** in the mineral clevite, the first discovery of the element on Earth's surface.

1896 French chemist Eugène-Anatole Demarçay discovers **europium**.

1898 English chemists Sir William Ramsay and Morris Travers discover **krypton, neon**, and **xenon**.

1898 French physicists Marie and Pierre Curie discover **polonium** and **radium**.

1898 German physicist Friedrich Ernst Dorn discovers **radon**.

1899 French chemist André Debierne discovers **actinium**.

1906 French chemist Georges Urbain and Austrian chemist Carl Auer von Welsbach independently discover **lutetium**.

1908 Japanese chemist Masataka Ogawa discovers **rhenium**, but erroneously assigns it to atomic number 43, instead of its correct atomic number of 75. He names the element nipponium, but his research is forgotten and ignored for many years.

1911 French chemist Georges Urbain and Russian chemist Vladimir Ivanovich Vernadskij independently discover **hafnium**. Their research is unconfirmed because of World War I (1914–1918).

1917 Three research teams, consisting of German physicists Lise Meitner and Otto Hahn, Polish-American physical chemist Kasimir Fajans and German chemist O. H. Göhring, and English physicists Frederick Soddy and John A. Cranston, independently and almost simultaneously discover **protactinium**.

1923 Dutch physicist Dirk Coster and Hungarian chemist George Charles de Hevesy rediscover **hafnium** and are generally recognized as discoverers of the element.

1925 German chemists Walter Noddack, Ida Tacke, and Otto Berg rediscover and name **rhenium**.

1937 Italian physicist Emilio Segré and his colleague Carlo Perrier discover **technetium**.

1939 French chemist Marguerite Perey discovers **francium**.

1940 Dale R. Corson, Kenneth R. Mackenzie, and Emilio Segré discover **astatine**.

1940 Edwin M. McMillan and Philip H. Abelson prepare **neptunium**.

1940 University of California at Berkeley researchers Glenn Seaborg, Arthur C. Wahl, J. K. Kennedy, and E. M. McMillan prepare **plutonium**.

1944 University of California at Berkeley researchers Glenn Seaborg, Albert Ghiorso, Ralph A. James, and Leon O. Morgan prepare **americium**.

1944 University of California at Berkeley researchers Glenn Seaborg, Albert Ghiorso, and Ralph A. James prepare **curium**.

1945 Oak Ridge National Laboratory researchers Jacob A. Marinsky, Lawrence E. Glendenin, and Charles D. Coryell discover **promethium**. The element had probably been found as early as 1942 by Chien Shiung Wu, Emilio Segré, and Hans Bethe.

1949 University of California at Berkeley researchers Stanley G. Thompson, Albert Ghiorso, and Glenn Seaborg prepare **berkelium**.

1950 University of California at Berkeley researchers Glenn Seaborg, Albert Ghiorso, Kenneth Street Jr., and Stanley G. Thompson prepare **californium**.

1952 A team of University of California at Berkeley researchers led by Albert Ghiorso prepares **einsteinium** (#99) and **fermium** (#100).

1955 University of California at Berkeley researchers Glenn Seaborg, Albert Ghiorso, Bernard G. Harvey, Gregory R. Choppin, and Stanley G. Thompson prepare **mendelevium** (#101).

1958 University of California at Berkeley researchers Glenn Seaborg, Albert Ghiorso, Tørbjorn Sikkeland, and J. R. Walton prepare **nobelium** (#102).

1961 University of California at Berkeley researchers Albert Ghiorso, Tørbjorn Sikkeland, Almon E. Larsh, and R. M. Latimer prepare **lawrencium** (#103).

1964 A research team at the Joint Institute for Nuclear Research at Dubna, Russia, led by Russian physicist Georgy Nikolaevich Flerov, produces **rutherfordium** (#104).

1968 A research team at the Joint Institute for Nuclear Research at Dubna, Russia, led by Russian physicist Georgy Nikolaevich Flerov, produces **dubnium** (#105).

1974 A research team at the University of California at Berkeley led by Albert Ghiorso produces **seaborgium** (#106).

1981 A research team led by Peter Armbruster and Gottfried Münzenberg at the Gesellschaft für Schwerionenforschung (Institute for Heavy Ion Research) in Darmstadt, Germany, produces **bohrium** (#107).

1982 The Heavy Ion Research team in Darmstadt produces **meitnerium** (#109).

1984 The Heavy Ion Research team in Darmstadt produces **hassium** (#108).

1994 The Heavy Ion Research team in Darmstadt, under the leadership of Sigurd Hofmann, produces **darmstadtium** (#110) and **roentgenium** (#111).

1996 The Heavy Ion Research team in Darmstadt, under the leadership of Sigurd Hofmann, produces **copernicium** (#112).

1999 Researchers at the Joint Institute for Nuclear Research announce the discovery of element #114, ununquadium.

2000 Researchers at the Joint Institute for Nuclear Research announce the discovery of element #116, ununhexium. This and all subsequent discoveries have not been confirmed nor a name approved for the possible new elements as of early 2010.

2002 Researchers at the Joint Institute for Nuclear Research at Dubna and the Lawrence Livermore National Laboratory in California announce the discovery of element #118, ununoctium.

2003 Researchers at the Joint Institute for Nuclear Research at Dubna and the Lawrence Livermore National Laboratory in California announce the discovery of element #113 ununtrium and element #115 ununpentium.

2008 Researchers at the Hebrew University at Jerusalem, under the leadership of Amnon Marinov, report the discovery of single atoms of element #122, unbibium, in naturally occurring deposits of **thorium**. Although unconfirmed as of early 2010, this report would be the first discovery of a new element in nature since 1939 (**francium**).

2009 The International Union of Pure and Applied Chemistry (IUPAC) officially recognizes element 112, Copernicium (Cn), and adds the name to the standard periodic table.

2009 A team of researchers at the Lawrence Berkeley National Laboratory confirm the production of #114, ununquadium.

Words to Know

Abrasive: A powdery material used to grind or polish other materials.

Absolute zero: The lowest temperature possible, about -459°F (-273°C).

Actinoid family: Formerly Actinide family; elements in the periodic table with atomic numbers 90 through 103.

Alchemy: A kind of pre-science that existed from about 500 BCE to about the end of the 16th century.

Alkali: A chemical with properties opposite those of an acid.

Alkali metal: An element in Group 1 (IA) of the periodic table.

Alkaline earth metal: An element found in Group 2 (IIA) of the periodic table.

Allotropes: Forms of an element with different physical and chemical properties.

Alloy: A mixture of two or more metals that has properties different from those of the individual metals.

Alpha particles: Tiny, atom-sized particles that can destroy cells.

Alpha radiation: A form of radiation that consists of very fast moving alpha particles and helium atoms without their electrons.

Amalgam: A combination of mercury and at least one other metal.

Amorphous: Without crystalline shape.

Anhydrous ammonia: Dry ammonia gas.

Antiseptic: A chemical that stops the growth of germs.

Aqua regia: A mixture of hydrochloric and nitric acids that often reacts with materials that do not react with either acid separately.

B

Battery: A device for changing chemical energy into electrical energy.

Biochemistry: The field of chemistry concerned with the study of compounds found in living organisms.

Biocompatible: Not causing a reaction when placed into the body.

Bipolar disorder: A condition in which a person experiences wild mood swings.

Brass: An alloy of copper and zinc.

Bronze: An alloy of copper and tin.

Bronze Age: A period in human history ranging from about 3500 to 1000 BCE, when bronze was widely used for weapons, utensils, and ornamental objects.

Buckminsterfullerene: Full name for buckyball or fullerene; also see Buckyball.

Buckyball: An allotrope of carbon whose 60 carbon atoms are arranged in a sphere-like form.

C

Capacitor: An electrical device, somewhat like a battery, that collects and then stores up electrical charges.

Carat: A unit of weight for gold and other precious metals, equal to one fifth of a gram, or 200 milligrams.

Carbon arc lamp: A lamp for producing very bright white light.

Carbon-14 dating: A technique that allows archaeologists to estimate the age of once-living materials by using the knowledge that carbon-14 is found in all living carbon materials; once an organism dies, no more carbon-14 remains.

Cassiterite: An ore of tin containing tin oxide, the major commercial source of tin metal.

Catalyst: A substance used to speed up a chemical reaction without undergoing any change itself.

Chalcogens: Elements in Group 16 (VIA) of the periodic table.

Chemical reagent: A substance, such as an acid or an alkali, used to study other substances.

Chlorofluorocarbons (CFCs): A family of chemical compounds consisting of carbon, fluorine, and chlorine that were once used widely as propellants in commercial sprays but regulated in the United States since 1987 because of their harmful environmental effects.

Corrosive agent: A material that tends to vigorously react or eat away at something.

Cyclotron: A particle accelerator, or "atom smasher," in which small particles, such as protons, are made to travel very fast and then collide with atoms, causing the atoms to break apart.

Density: The mass of a substance per unit volume.

Diagnosis: Finding out what medical problems a person may have.

Distillation: A process by which two or more liquids can be separated from each other by heating them to their boiling points.

"Doped": Containing a small amount of a material as an impurity.

Ductile: Capable of being drawn into thin wires.

Earth: In mineralogy, a naturally occurring form of an element, often an oxide of the element.

Electrolysis: A process by which a compound is broken down by passing an electric current through it.

Electroplating: The process by which a thin layer of one metal is laid down on top of a second metal.

Enzyme: A substance that stimulates certain chemical reactions in the body.

Fabrication: Shaping, molding, bending, cutting, and working with a metal.

Fission: The process by which large atoms break apart, releasing large amounts of energy, smaller atoms, and neutrons in the process.

Fly ash: The powdery material produced during the production of iron or some other metal.

Frasch method: A method for removing sulfur from underground mines by pumping hot air and water down a set of pipes.

Fuel cell: Any system that uses chemical reactions to produce electricity.

Fullerene: Alternative name for buckyball; also see Buckyball.

Galvanizing: The process of laying down a thin layer of zinc on the surface of a second metal.

Gamma rays: A form of radiation similar to X rays.

Global warming: A phenomenon in which the average temperature of Earth rises, melting icecaps, raising sea levels, and causing other environmental problems. Causes include human-activities, including heavy emissions of carbon dioxide (CO_2).

Half life: The time it takes for half of a sample of a radioactive element to break down.

Halogen: One of the elements in Group 17 (VIIA) of the periodic table.

Heat exchange medium: A material that picks up heat in one place and carries it to another place.

Hydrocarbons: Compounds made of carbon and hydrogen.

Hypoallergenic: Not causing an allergic reaction.

Inactive: Does not react with any other element.

Inert: Not very active.

Isotope: Two or more forms of an element that differ from each other according to their mass number.

Lanthanoid family: Formerly Lanthanide family; the elements in the periodic table with atomic numbers 57 through 71.

Laser: A device for making very intense light of one very specific color that is intensified many times over.

Liquid air: Air that has been cooled to a very low temperature.

Luminescence: The property of giving off light without giving off heat.

Machining: The bending, cutting, and shaping of a metal by mechanical means.

"Magic number": The number of protons and/or neutrons in an atom that tend to make the atom stable (not radioactive).

Magnetic field: The space around an electric current or a magnet in which a magnetic force can be observed.

Malleable: Capable of being hammered into thin sheets.

Metalloid: An element that has characteristics of both metals and non-metals.

Metallurgy: The art and science of working with metals.

Metals: Elements that have a shiny surface, are good conductors of heat and electricity, and can be melted, hammered into thin sheets, and drawn into thin wires.

Micronutrient: A substance needed in very small amounts to maintain good health.

Misch metal: A metal that contains different rare earth elements and has the unusual property of giving off a spark when struck.

Mohs scale: A way of expressing the hardness of a material.

Mordant: A material that helps a dye stick to cloth.

Nanotubes: Long, thin, and extremely tiny tubes.

Native: Not combined with any other element.

Neutron radiography: A technique that uses neutrons to study the internal composition of material.

Nickel allergy: A health condition caused by exposure to nickel metal.

Nitrogen fixation: The process of converting nitrogen as an element to a compound that contains nitrogen.

Noble gases: Elements in Group 18 (VIIIA) of the periodic table.

Non-metals: Elements that do not have the properties of metals.

Nuclear fission: A process in which neutrons collide with the nucleus of a plutonium or uranium atom, causing it to split apart with the release of very large amounts of energy.

Nuclear reactor: A device in which nuclear reactions occur.

Optical fiber: A thin strand of glass through which light passes; the light carries a message through a telephone wire.

Ore: A mineral compound that is mined for one of the elements it contains, usually a metal element.

Organic chemistry: The study of the carbon compounds.

Oxidizing agent: A chemical substance that gives up or takes on electrons from another substance.

Ozone: A form of oxygen that filters out harmful radiation from the sun.

Ozone layer: The layer of ozone that shields Earth from harmful ultraviolet radiation from the sun.

Particle accelerator ("atom smasher"): A device used to cause small particles, such as protons, to move at very high speeds.

Periodic law: A way of organizing the chemical elements to show how they are related to each other.

Periodic table: A chart that shows how chemical elements are related to each other.

Phosphor: A material that gives off light when struck by electrons.

Photosynthesis: The process by which plants convert carbon dioxide and water to carbohydrates (starches and sugars).

Platinum family: A group of elements that occur close to platinum in the periodic table and with platinum in the Earth's surface.

Polymerization: The process by which many thousands of individual tetrafluoroethlylene (TFE) molecules join together to make one very large molecule.

Potash: A potassium compound that forms when wood burns.

Precious metal: A metal that is rare, desirable, and, therefore, expensive.

Proteins: Compounds that are vital to the building and growth of cells.

Pyrophoric: Gives off sparks when scratched.

Radiation: Energy transmitted in the form of electromagnetic waves or subatomic particles.

Radioactive isotope: An isotope that breaks apart and gives off some form of radiation.

Radioactive tracer: An isotope whose movement in the body can be followed because of the radiation it gives off.

Radioactivity: The process by which an isotope element breaks down and gives off some form of radiation.

Rare earth elements: Elements in the Lanthanoid family.

Reactive: Combines with other substances relatively easily.

Refractory: A material that can withstand very high temperatures and reflects heat back away from itself.

Rodenticide: A poison used to kill rats and mice.

Rusting: A process by which a metal combines with oxygen.

Salt dome: A large mass of salt found underground.

Semiconductor: A material that conducts an electric current, but not nearly as well as metals.

Serendipity: Discovering something of value when not seeking it; for example, making a discovery by chance or accident.

Silver plating: A process by which a very thin layer of silver metal is laid down on top of another metal.

Slag: A mixture of materials that separates from a metal during its purification and floats on top of the molten metal.

Slurry: A soup-like mixture of crushed ore and water.

Solder: An alloy that can be melted and then used to join two metals to each other.

Spectra: The lines produced when chemical elements are heated.

Spectroscope: A device for analyzing the light produced when an element is heated.

Spectroscopy: The process of analyzing light produced when an element is heated.

Spectrum (plural: spectra): The pattern of light given off by a glowing object, such as a star.

Stable: Not likely to react with other materials.

Sublimation: The process by which a solid changes directly to a gas when heated, without first changing to a liquid.

Superalloy: An alloy made of iron, cobalt, or nickel that has special properties, such as the ability to withstand high temperatures and attack by oxygen.

Superconductivity: The tendency of an electric current to flow through a material without resistance.

Superconductor: A material that has no resistance to the flow of electricity; once an electrical current begins flowing in the material, it continues to flow forever.

Superheated water: Water that is hotter than its boiling point, but which has not started to boil.

Surface tension: A property of liquids that makes them act like they are covered with a skin.

Tarnishing: Oxidizing; reacting with oxygen in the air.

Tensile: Capable of being stretched without breaking.

Thermocouple: A device for measuring very high temperatures.

Tin cry: A screeching-like sound made when tin metal is bent.

Tin disease: A change that takes place in materials containing tin when the material is cooled to temperatures below 55°F (13°C) for long periods of time, when solid tin turns to a crumbly powder.

Tincture: A solution made by dissolving a substance in alcohol.

Tinplate: A type of metal consisting of a thin protective coating of tin deposited on the outer surface of some other metal.

Toxic: Poisonous.

Trace element: An element that is needed in very small amounts for the proper growth of a plant or animal.

Tracer: A radioactive isotope whose presence in a system can easily be detected.

Transfermium element: Any element with an atomic number greater than 100.

Transistor: A device used to control the flow of electricity in a circuit.

Transition metal: An element in Groups 3 through 12 of the periodic table.

Transuranium element: An element with an atomic number greater than 92.

Ultraviolet (UV) radiation: Electromagnetic radiation (energy) of a wavelength just shorter than the violet (shortest wavelength) end of the visible light spectrum and thus with higher energy than visible light.

Vulcanizing: The process by which soft rubber is converted to a harder, long-lasting product.

Workability: The ability to work with a metal to get it into a desired shape or thickness.

Gadolinium

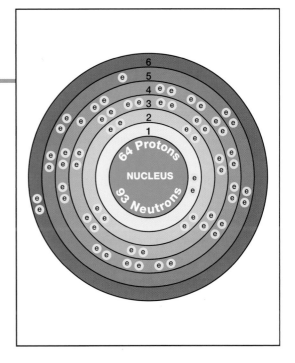

Overview

Gadolinium was named for Finnish chemist Johan Gadolin (1760–1852). Gadolin served for many years as professor of chemistry at the University of Åbo in Finland. He was the first person to study an unusual black stone discovered near the town of Ytterby, Sweden, in 1787. The stone was an unusually important discovery. Chemists isolated nine new elements from the stone, one of which was gadolinium.

Gadolinium is in Row 6 of the periodic table. The periodic table is a chart that shows how chemical elements are related to each other. It is one of the rare earth elements. They really aren't rare, but they *were* difficult to separate. For this reason, scientists know less about the rare earth elements than they do about most other elements. The rare earth metals are also called lanthanoids. That name comes from element 57, **lanthanum**.

Discovery and Naming

Two unusual rocks were discovered in Sweden near the end of the 18th century. The rocks were unusual because they both contained a complex mixture of substances. Chemists worked for nearly a century to separate

Key Facts

Symbol: Gd

Atomic Number: 64

Atomic Mass: 157.25

Family: Lanthanoid (rare earth metal)

Pronunciation: gad-uh-LIN-ee-um

WORDS TO KNOW

Ductile: Capable of being drawn into a thin wire.

Isotopes: Two or more forms of an element that differ from each other according to their mass number.

Lanthanoids (rare earth elements): The elements in the periodic table with atomic numbers 57 through 71.

Malleable: Capable of being hammered into thin sheets.

Neutron radiography: A technique that uses neutrons to study the internal composition of material.

Nuclear fission: The process in which large atoms break apart, releasing large amounts of energy, smaller atoms, and neutrons in the process.

Periodic table: A chart that shows how the chemical elements are related to each other.

Phosphor: A material that gives off light when struck by electrons.

Radioactive isotope: An isotope that breaks apart and gives off some form of radiation.

Toxic: Poisonous.

the mixtures and find out what they were. The 15 rare earth elements were first discovered in the two Swedish rocks.

One rock contained a mineral that had never been seen before, cerite. Cerite was first discovered in 1803. The last new element found in cerite was not identified until almost a century later, in 1901. In 1880, French chemist Jean-Charles-Galissard de Marignac (1817–1894) was studying a new material found in cerite called samaria. Earlier chemists had identified samaria as a new element.

Marignac found that samaria was not a pure element. Instead, it consisted of two parts, which he called samaria and gadolinia. He believed each was a new element. He was right about gadolinia, but wrong about samaria.

Physical Properties

Gadolinium has a shiny metallic luster with a slight yellowish tint. It is both ductile and malleable. Ductile means capable of being made into wires. Malleable means capable of being hammered or rolled into thin sheets. It has a melting point of 2,394°F (1,312°C) and a boiling point of about 5,400°F (3,000°C). Its density is 7.87 grams per cubic centimeter.

Few elements are as strongly magnetic as gadolinium. It also has the highest neutron-absorbing ability of any element. A piece of gadolinium stops neutrons better than any other element.

What's in a Name?

Books on the chemical elements sometimes talk about gadolin*ium* and gadolin*ia*; about erb*ium* and sometimes erb*ia*; sometimes about samar*ium* and sometimes samar*ia*. Is the *-ium* ending any different from the *-ia* ending?

The answer is yes. Almost all metal names end in *-ium* or just *-um*, like sod*ium*, potass*ium*, magnes*ium*, alumin*um*, and gadolin*ium*. The ending *-ia* or just *-a*, on the other hand, stands for the form in which an element occurs in the earth. When miners take gadolinium out of the earth, it is called *gadolinia*.

The natural form of the element is often called an "earth." Gadolinium is the element that comes from the earth, gadolinia.

Earths are compounds of the element and one or more other element. Two common combining elements are oxygen and sulfur. For example, gadolinia contains gadolinium oxide (Gd_2O_3).

These terms can be confusing when reading the history of chemical elements. Many elements were first discovered not in their pure form, but as compounds—as earths.

Chemical Properties

Gadolinium metal is not especially reactive. It dissolves in acids and reacts slowly with cold water. It also reacts with oxygen at high temperatures.

Occurrence in Nature

The abundance of gadolinium in Earth's surface is estimated at about 4.5 to 6.4 parts per million. That would make it one of the most abundant of the rare earth elements. It ranks above **bromine** and **uranium**, but just below **lead** and **boron** in order of abundance. Some minerals in which it occurs are monazite, bastnasite, samarskite, gadolinite, and xenotime.

Isotopes

Seven naturally occurring isotopes of gadolinium are known. They are gadolinium-152, gadolinium-154, gadolinium-155, gadolinium-156, gadolinium-157, gadolinium-158, and gadolinium-160. Isotopes are two or more forms of an element. Isotopes differ from each other according to their mass number. The number written to the right of the element's name is the mass number. The mass number represents the number of protons plus neutrons in the nucleus of an atom of the element. The number of protons determines the element, but the number of neutrons in the atom of any one element can vary. Each variation is an isotope. One of gadolinium's naturally occurring isotopes is radioactive, gadolinium-152. A radioactive isotope is one that breaks apart and gives off some form of radiation.

Radioactive isotopes can also be produced artificially when very small particles are fired at atoms. These particles stick in the atoms and make them radioactive. Twenty-four artificial radioactive isotopes have been produced. Some of these are used in medicine. For example, gadolinium-153 is used to study the composition of bones. The radiation it gives off acts like X rays, penetrating the bones to reveal the minerals present.

Gadolinium is also used in another X-ray-like technique called neutron radiography. In this technique, neutrons are fired through a sample of material. The neutrons act somewhat like X rays. They show the structure of the material. Adding gadolinium to the back side of the material makes the neutron image easier to read. Neutron radiography is especially useful because one can look for damage inside a piece of metal without having to take the material apart.

Extraction

The method for obtaining gadolinium from its ores is similar to that for other rare earth elements. The ore is converted into gadolinium chloride ($GdCl_3$) or gadolinium fluoride (GdF_3). Passing an electric current through the first compound releases pure gadolinium:

$$2GdCl_3 \text{ —electric current} \rightarrow 2Gd + 3Cl_2$$

Adding calcium to the second compound also releases pure gadolinium:

$$3Ca + 2GdF_3 \rightarrow 3CaF_2 + 2Gd$$

Uses and Compounds

Gadolinium is used in control rods in nuclear power plants. Energy produced during nuclear fission is used to generate electricity. Nuclear fission is the process in which large atoms (usually uranium or **plutonium**) break apart, releasing energy. The smaller atoms produced are called fission products and are radioactive.

Neutrons are also produced in the reaction. In order for a nuclear power plant to work properly, the number of neutrons must be carefully controlled. Rods containing gadolinium are raised out of or lowered into the reactor. This allows more or fewer neutrons to remain in the reaction.

Gadolinium also has medical uses. It is used to locate the presence of tumors in the inner ear. Gadolinium is injected into the bloodstream. It then goes to any tumor that happens to be present in the ear. The tumor appears darker when seen with X rays.

Gadolinium compounds are used as phosphors in television tubes. A phosphor is a material that shines when struck by electrons. The color of the phosphor depends on the elements of which it is made.

Gadolinium is also found in alloys and special minerals known as **yttrium** garnets. An alloy is made by melting and mixing two or more metals. The mixture has properties different from those of the individual metals. Gadolinium alloys are easier to work with than alloys without gadolinium. Gadolinium yttrium garnets are used in microwave ovens to produce the microwaves.

Health Effects

Not many details of the health effects of gadolinium are known. It is usually handled as if it were very toxic.

Gallium

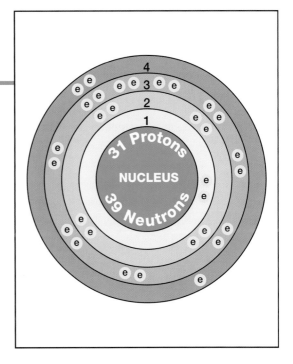

Overview

In the late 1860s Russian chemist Dmitri Mendeleev (1834–1907) made one of the greatest discoveries in modern chemistry: the periodic law. The periodic law describes how chemical elements are related to each other. The periodic law is often depicted in the form of the periodic table. The periodic table is a chart that lists all of the chemical elements and sorts them into groups based on similarities. Elements in vertical columns are similar to each other in many ways.

When Mendeleev first proposed the periodic law, he made a troubling discovery. There were a few empty spots in his table. For example, the box set aside for element number 31 was empty. No element had been found that belonged in that box.

Part of Mendeleev's genius was what he did next. He said that an "element number 31" did exist. Scientists simply had not found it yet. Mendeleev then described what the element would be like. He based his prediction on elements on all sides of the box for element number 31. He said it would be similar to **aluminum** (in box 13, above 31) and **indium** (in box 49, below 31). He named this missing element eka-aluminum.

Key Facts

Symbol: Ga

Atomic Number: 31

Atomic Mass: 69.723

Family: Group 13 (IIIA); aluminum

Pronunciation: GA-le-um

WORDS TO KNOW

Ductile: Capable of being drawn into thin wires.

Isotopes: Two or more forms of an element that differ from each other according to their mass number.

Laser: A device for making very intense light of one very specific color that is intensified many times over.

Periodic table: A chart that shows how chemical elements are related to each other.

Radioactive isotope: An isotope that breaks apart and gives off some form of radiation.

Semiconductor: A material that conducts an electrical current.

Spectra: The lines produced when chemical elements are heated.

Transistor: A device used to control the flow of electricity in a circuit.

Using Mendeleev's periodic law, the element was soon found. It was discovered by French chemist Paul-émile Lecoq de Boisbaudran in 1875.

Until recently, gallium had few applications. Then, some of its compounds were discovered to have unusual properties when exposed to light. These properties make gallium an important and essential element in many electronic devices.

Discovery and Naming

Lecoq de Boisbaudran did not discover gallium by accident. For 15 years, he had studied the spectra of the chemical elements. Spectra (singular: spectrum) are the lines produced when chemical elements are heated. Each element produces its own distinctive set of lines, or spectra. An element can be identified in a sample by the spectrum it produces.

Lecoq de Boisbaudran knew that the element between aluminum and indium was missing. He also knew about Mendeleev's prediction. Lecoq de Boisbaudran wanted to learn more about the spectra of elements. He thought that element number 31 might be found in **zinc** ores. Zinc has an atomic number of 30, so it is next to gallium on the periodic table.

Lecoq de Boisbaudran had to work through a large amount of zinc ore. But his hunch turned out to be correct. The missing element was present in the ore, but only in very small amounts. Finally, in August 1875, Lecoq de Boisbaudran reported that "the new substance gave under the action of the electric spark a spectrum composed chiefly of a violet ray, narrow, readily visible, and [located at] about 417 on the scale of wave lengths."

Later in the same year, Lecoq de Boisbaudran isolated gallium metal. He was given several tons of zinc ore by miners for his research. Out of this ore, he was able to produce a few grams of nearly pure gallium.

Lecoq de Boisbaudran proposed the name gallium for the new element. The name was given in honor of the ancient name for France, Gallia.

Physical Properties

Gallium is a soft, silvery metal with a shiny surface. In some ways, however, it is very un-metal-like. It is so soft that it can be cut with a knife. It has a very low melting point of only 85.5°F (29.7°C). A sample of gallium will melt if held in the human hand (body temperature, about 98.6°F/37°C).

Another unusual property is that gallium can be supercooled rather easily. Supercooling is the cooling of a substance below its freezing point without it becoming a solid. Gallium is a liquid at 86°F (30°C), so one would expect it to become a solid at 85.5°F (29.7°C). Instead it is fairly easy to cool gallium to below 85.5°F (29.7°C) without having it solidify.

Gallium melts when held in one's hand. © RUSS LAPPA/ SCIENCE SOURCE, NATIONAL AUDUBON SOCIETY COLLECTION/PHOTO RESEARCHERS, INC.

Gallium's boiling point is about 4,400°F (2,400°C) and its density is 5.9037 grams per cubic centimeter, almost six times the density of water.

Chemical Properties

Gallium is a fairly reactive element. It combines with most metals at high temperatures, and it reacts with both acids and alkalis. An alkali is a chemical with properties opposite those of an acid. Sodium hydroxide (common lye) is an example of an alkali.

Occurrence in Nature

Gallium is a moderately abundant element in Earth's crust. Its abundance is estimated to be about 5 parts per million. It is found primarily in combination with zinc and aluminum ores. It is also found in germanite, an ore of **copper** sulfide (CuS).

As of 2008, the largest producers of gallium were China, Germany, Kazakhstan, and Ukraine. According to the U.S. Geological Survey (USGS), leaders in refined gallium production included China, Japan, and the United States.

Isotopes

Two naturally occurring isotopes of gallium are known: gallium-69 and gallium-71. Isotopes are two or more forms of an element. Isotopes differ from each other according to their mass number. The number written to the right of the element's name is the mass number. The mass number represents the number of protons plus neutrons in the nucleus of an atom of the element. The number of protons determines the element, but the number of neutrons in the atom of any one element can vary. Each variation is an isotope.

Twenty-four radioactive isotopes of gallium are known also. A radioactive isotope is one that breaks apart and gives off some form of radiation. Radioactive isotopes are produced when very small particles are fired at atoms. These particles stick in the atoms and make them radioactive.

One radioactive isotope of gallium, gallium-67, has long been used in medicine. This isotope has a tendency to seek out cancer cells in the body. Its presence in a cell can be detected by the radiation it gives off. By giving a patient a dose of gallium-67, a doctor can determine whether the patient has cancer. Gallium-67 has been used to look for cancer in the liver, spleen, bowels, breasts, thymus, kidneys, and bones.

Extraction

Pure gallium metal can be prepared by passing an electric current through a gallium compound, such as gallium oxide (Ga_2O_3):

$$2Ga_2O_3 \xrightarrow{\text{electric current}} 4Ga + 3O_2$$

Uses and Compounds

About 95 percent of all gallium produced is used to make a single compound, gallium arsenide (GaAs). Gallium arsenide has the ability to convert an electrical current directly into light. The lighted numbers on hand-held calculators, for example, are produced by a device known as a light-emitting diode (LED). Gallium arsenide is used to make LEDs.

Close-up view of LED numerical display on a digital scale. Gallium arsenide is used to make LEDs. IMAGE COPYRIGHT 2009, BALONCICI. USED UNDER LICENSE FROM SHUTTERSTOCK.COM.

An LED allows an electric current to flow in one side, but not the other. When it flows into a piece of gallium arsenide, a flash of light is produced. When a button is pushed on a calculator, a circuit is closed. The electric current flows into an LED and produces a light.

Similar devices are used in making lasers. An electric current passes into a piece of gallium arsenide. The current produces an intense beam of laser light. A laser is a device for producing very bright light of a single color. Gallium arsenide lasers are used in a number of applications. The laser that operates a compact disc (CD) player, for example, may contain a piece of gallium arsenide.

Gallium arsenide is also used to make transistors. A transistor is a device used to control the flow of electricity in a circuit. Gallium arsenide has many of the properties of a semiconductor. A semiconductor is a material that conducts an electrical current, but not as well as a metal, such as **silver** or copper.

Gallium arsenide has one big advantage over **silicon**, another element used in transistors. Gallium arsenide produces less heat. Therefore,

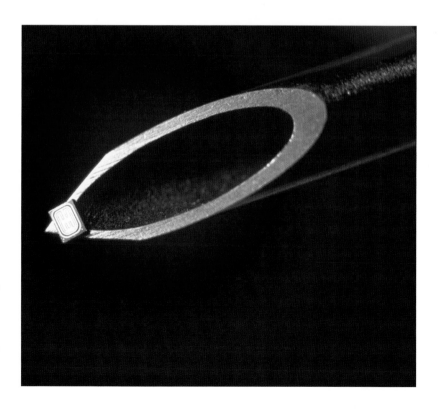

Gallium arsenide is used in the production of transistors.
© MICHAEL ABBEY. SCIENCE SOURCE, THE NATIONAL AUDUBON SOCIETY COLLECTION/PHOTO RESEARCHERS, INC.

more transistors can work together at the same time to produce a higher computing capacity.

Gallium arsenide is also used in photovoltaic cells. These devices turn sunlight into electricity. Many people believe that photovoltaic cells will someday replace coal-fired generating and nuclear power plants as the major source of electricity.

Large amounts of gallium have also been used to study radiation given off by the sun. In experiments conducted at the National Laboratory at Gran Sasso, Italy, containers holding more than 33 short tons (30 metric tons) of gallium were used to detect and measure various types of radiation originating from the sun. These experiments were conducted from 1991 to 1997 and from 1998 to 2002.

Health Effects

Gallium and its compounds are somewhat hazardous to the health of humans and animals. They produce a metallic taste in the mouth, skin rash, and a decrease in the production of blood cells. Gallium and its compounds should be handled with caution.

Germanium

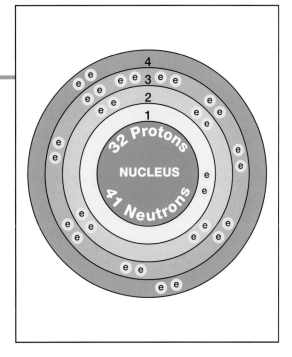

Overview

Germanium is a metalloid. A metalloid is an element that has characteristics of both metals and non-metals. Germanium is located in the middle of the **carbon** family, which is Group 14 (IVA) in the periodic table. The periodic table is a chart that shows how chemical elements are related to each other. Carbon and **silicon** are above germanium and **tin** and **lead** are below it.

The existence of germanium was predicted by Russian chemist Dmitri Mendeleev (1834–1907), who developed the periodic table. Mendeleev predicted a new element would be found to fill an empty spot on the table. He was proved correct in 1886.

Until the 1950s, there were no important uses for germanium. Then, the transistor was invented. A transistor is a device used to control the flow of electricity in a circuit. Today, germanium is used in making a number of electronic devices, including the transistor.

Discovery and Naming

In the 1860s, Mendeleev wondered if the chemical elements could be arranged in any systematic way. Are all chemical elements very different from each other, he asked. Do they have certain common properties?

Key Facts

Symbol: Ge

Atomic Number: 32

Atomic Mass: 72.64

Family: Group 14 (IVA); carbon

Pronunciation: jur-MAY-nee-um

WORDS TO KNOW

Catalyst: A substance used to speed up a chemical reaction without undergoing any change itself.

Doping: Containing a small amount of a material as an impurity.

Isotopes: Two or more forms of an element that differ from each other according to their mass number.

Metalloid: An element that has characteristics of both metals and non-metals.

Periodic table: A chart that shows how the chemical elements are related to each other.

Radioactive isotope: An isotope that breaks apart and gives off some form of radiation.

Semiconductor: A material that conducts an electric current, but not nearly as well as metals.

Transistor: A device used to control the flow of electricity in a circuit.

He explored a number of ways of arranging the elements. Finally, he decided to arrange them according to their atomic weights. He found that doing so resulted in a pattern. After awhile, each element could be placed in a position beneath one or more elements before it. Mendeleev described his result in the periodic law. The periodic table is the most common way of illustrating the periodic law. Elements that are similar to each other fall into the same group. For example, the elements in Group 1 (IA) are like each other in many ways.

Mendeleev found that his periodic table made sense, however, only if he left some blank spaces in it. His table had a blank space for element number 32. No element was known to exist at that time that had properties like silicon (number 14) and could be put beneath it in the periodic table.

This finding did not disturb Mendeleev. Element number 32 simply had not been discovered yet, he said.

A number of chemists took up Mendeleev's challenge. In 1885, a new ore was discovered in a mine near Freiberg, Germany. **Silver** and **sulfur** were found in the ore, but about 7 percent of the ore could not be identified. The ore was sent to German chemist Clemens Alexander Winkler (1838–1904). At the time, Winkler was professor of chemical technology and analytical chemistry at the Freiberg School of Mines. He was convinced that the new ore contained a new element. He isolated the new element from the ore and named it germanium, in honor of Winkler's native country, Germany.

Winkler discovered that the properties of germanium were very similar to those that Mendeleev had predicted 15 years earlier. For example, he thought element 32 would have a density of 5.5 grams per cubic centimeter. The actual density for germanium is 5.47 grams per cubic centimeter. Mendeleev had based his predictions on the new element's place in the periodic table. His success in making these predictions gave chemists a great deal of confidence in the periodic table. They came to see how useful it could be in their research.

Physical Properties

Germanium looks like a metal. It has a bright, shiny, silvery color. But it is brittle and breaks apart rather easily, which metals normally do not do. It has a melting point of 1,719°F (937.4°C) and a boiling point of 5,130°F (2,830°C). It conducts an electric current poorly. Substances of this kind are called semiconductors, which conduct an electric current, but not nearly as well as metals like silver, **copper**, and **aluminum**.

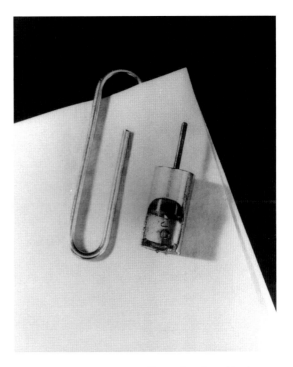

Germanium is used in the production of transistors. This transistor was made in the 1950s. LIBRARY OF CONGRESS.

The ability of semiconductors to conduct electricity depends greatly on the presence of small amounts of impurities. The addition of an impurity to a semiconductor is called doping. Doping a semiconductor has significant effects on its ability to conduct an electric current.

Chemical Properties

Germanium is a relatively inactive element. It does not react with **oxygen** at room temperature. It does dissolve in hot acids and with oxygen at high temperatures, however. It becomes more active when finely divided. It will combine with **chlorine** and **bromine** to form germanium chloride ($GeCl_4$) and germanium bromide ($GeBr_4$). For example:

$$Ge \text{ (finely divided)} + 2Cl_2 \xrightarrow{\text{high temperature}} GeCl_4$$

Occurrence in Nature

The abundance of germanium in Earth's crust is estimated to be about 7 parts per million. That places it in the bottom third of the elements arranged according to their abundance.

The two most common minerals of germanium are argyrodite and germanite. Argyrodite is the mineral in which Winkler first discovered germanium. Germanite contains about 8 percent germanium.

Most germanium today is obtained from **zinc** ores. When those ores are treated to obtain zinc metal, some germanium is produced at the same time.

In 2008, germanium from zinc ores was obtained from several mines in the United States: one mine in Alaska, one in Washington state, and another in Tennessee. According to the U.S. Geological Survey (USGS), germanium refineries were in operation in New York and Oklahoma. The United States also imports germanium from Belgium, Canada, China, Germany, and a few other nations.

Isotopes

There are five naturally occurring isotopes of germanium: germanium-70, germanium-72, germanium-73, germanium-74, and germanium-76. Isotopes are two or more forms of an element. Isotopes differ from each other according to their mass number. The number written to the right of the element's name is the mass number. The mass number represents the number of protons plus neutrons in the nucleus of an atom of the element. The number of protons determines the element, but the number of neutrons in the atom of any one element can vary. Each variation is an isotope.

Twenty-five radioactive isotopes of germanium are known also. A radioactive isotope is one that breaks apart and gives off some form of radiation. Radioactive isotopes are produced when very small particles are fired at atoms. These particles stick in the atoms and make them radioactive.

None of the radioactive isotopes of germanium has any important commercial use.

Extraction

Germanium in zinc ores is heated in the presence of chlorine gas. Germanium chloride ($GeCl_4$) is formed:

$$Ge \text{ (in ores)} + 2Cl_2 \xrightarrow{\text{heat}} GeCl_4$$

Pure germanium metal is then produced by passing an electric current through molten (melted) germanium chloride:

$$GeCl_4 \xrightarrow{\text{electric current}} Ge + 2Cl_2$$

This method produces very pure germanium. It is possible to buy germanium that is 99.9999 percent pure. This level of purity is needed in order to use the metal in the production of semiconductors.

Uses

Germanium first became important for its use in semiconductors. This application still accounts for about 15 percent of the germanium produced. But other uses of the element are now more important. About 50 percent of the germanium produced in the United States is now used in the manufacture of infrared optics and another 30 percent in fiber optic systems.

An optical fiber is a very thin thread made out of pure glass. The fiber acts somewhat like a copper wire. It can carry messages in the same way a copper wire carries messages. The difference is that optical fibers carry messages on light waves. Copper wires carry messages on electric currents.

The ability of a glass thread to carry light depends on the presence of impurities. Optical fibers are doped with germanium and other elements to improve their ability to carry light messages. Optical fibers are used to carry telephone messages instead of electric wires.

Germanium is also used as a catalyst. A catalyst is used to speed up a chemical reaction. The catalyst does not undergo any chemical change during the reaction. Germanium catalysts are used primarily in the production of plastics.

Germanium is also used to make specialized glass for military applications. For example, it is used to make weapons-sighting systems that can be used in the dark as well as other night

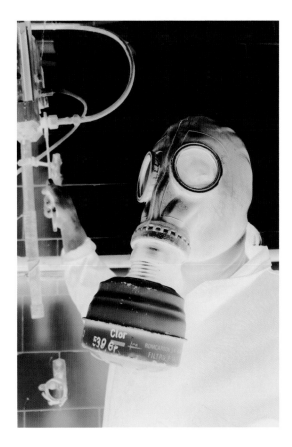

Germanium is used in night vision equipment. Here, a man wearing a gas mask becomes visible at night to others who are using night vision technology. IMAGE COPYRIGHT 2009, STUDIO 37. USED UNDER LICENSE FROM SHUTTERSTOCK. COM.

vision technology. Satellite systems and fire alarm systems may also contain glass that contains germanium.

Compounds

Few germanium compounds have any important commercial uses.

Health Effects

Germanium is not thought to be essential to the health of plants or animals. Some of its compounds present a hazard to human health, however. For example, germanium chloride and germanium fluoride (GeF_4) are a liquid and gas, respectively, that can be very irritating to the eyes, skin, lungs, and throat.

Gold

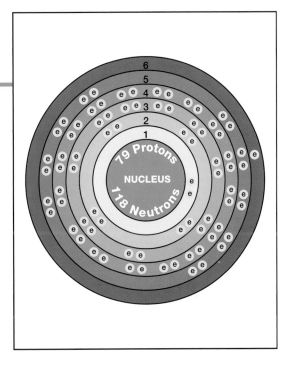

Overview

Gold has been called the most beautiful of all chemical elements. Its beauty has made it desirable for use in jewelry, coins, and artwork for thousands of years. It was one of the first pure metals to be used by humans.

Gold is one of the few elements that can affect politics and economics. Wars have been fought over access to gold. Cities and towns have sprung up and died out as gold was discovered and then mined out. Many nations still count their wealth according to the amount of gold they keep in storage.

Gold lies in the middle of the periodic table. The periodic table is a chart that shows how elements are related to one another. Gold is a heavy metal in a group known as the transition metals. Gold is also known as a precious metal (as are **platinum** and **silver**).

Large amounts of gold are still used in the manufacture of coins, medals, jewelry, and art. Gold also has a number of uses in industry, medicine, and other applications. For example, one radioactive isotope of gold is commonly used to treat cancer.

The chemical symbol for gold is Au. The symbol comes from the Latin word for gold, *aurum*. Aurum means "shining dawn."

Key Facts

Symbol: Au

Atomic Number: 79

Atomic Mass: 196.966569

Family: Group 11 (IB); transition metal

Pronunciation: GOLD

WORDS TO KNOW

Alloy: A mixture of two or more metals with properties different from those of the individual metals.

Amalgam: A mixture of two or more metals, one of which is mercury.

Carat: A unit of weight for gold and other precious metals, equal to one fifth of a gram, or 200 milligrams.

Ductile: Capable of being drawn into thin wires.

Isotopes: Two or more forms of an element that differ from each other according to their mass number.

Malleable: Capable of being hammered into thin sheets.

Radioactive isotope: An isotope that breaks apart and gives off some form of radiation.

Transition metal: An element in Groups 3 through 12 of the periodic table.

Discovery and Naming

Gold objects dating to 2600 BCE have been found. They were discovered in the royal tombs of the ancient civilization of Ur, located in the modern country of Iraq. These objects showed that humans had already learned how to work with gold this early in history. Some of the gold, for example, had been formed into wires.

One of the special skills developed by the Egyptians was adding gold to glass objects. They found a way to use gold to make glass a beautiful ruby-red color. The glass became known as gold ruby glass.

Gold is also mentioned in a number of places in the Bible. A passage in Exodus, for example, refers to the clothing worn by Aaron: "And they did beat the gold into thin plates, and cut it into wires, to work it in the blue, and in the purple, and in the scarlet, and in the fine linen, with cunning work."

Writings from every stage of human history tell of the discovery and use of gold. Roman historian Pliny the Elder (23–79 CE), for example, describes gold-mining locations. The Romans found gold lying in stream beds in the Tagus River in Spain, the Po River in Italy, the Hebrus River in Thracia (now Greece), the Pactolus River in Asia Minor (now Turkey), and the Ganges River in India.

Gold has long been known in the New World, too. During a visit to Haiti, Christopher Columbus (1451–1506) found gold nuggets lying on the bottom of rivers and harbors. A Portuguese explorer in 1586,

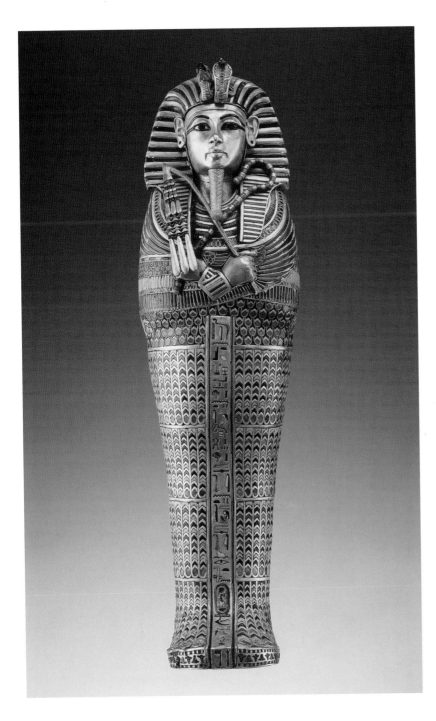

The coffin of Egyptian Pharaoh Tutankhamun (1358–1340 BCE) is inlaid with gold and other materials. AP IMAGES/ MUSEUM OF ANTIQUITIES BASEL, VOEGELIN.

Lopez Vaz, wrote that the region called Veragua (now Panama) was the "richest Land of Gold [in] all the rest of the Indies."

The Gold Rush As early as the 16th century, records contained stories about a great El Dorado ("the gilded one," in Spanish; gilded means "covered in gold") on the western coast of the United States. Tales of this magical city were repeated for centuries.

In the late 1840s, explorers began to travel from the Eastern seaboard to California in search of El Dorado. The flow of visitors was slow at first. Then, gold was first discovered in 1848 at a place called Sutter's Mill in California. Sutter's Mill is located near the present town of Coloma.

Word of the discovery spread quickly. Within a year, thousands of men and women made the long, expensive, and tiring trip. Many of them traveled across the United States in covered wagons or on horseback. Most had to cross mountains, plains, and deserts. Because of the difficult conditions, many people and animals got sick or died. Other people

Gold was discovered in California in January of 1848, prompting thousands to head West in a quest for fortune. By year's end, some $6 million in gold dust and nuggets had been retrieved. LIBRARY OF CONGRESS.

braved the seas and sailed to California, heading around Cape Horn at the bottom of South America or across the Isthmus of Panama. No matter which route was used, the journey usually took months.

As people arrived in California, hundreds of mining camps sprang up. Some of them had colorful names. Poker Flat, Hangtown, Red Dog, Hell's Delight, and Whiskey Bar were just a few! Mining for gold was hard work. Gold miners usually wound up being wildly successful or terrible failures. The Gold Rush completely changed the state of California. It also helped expand the United States.

Physical Properties

Gold is both ductile and malleable. Ductile means it can be drawn into thin wires. Malleable means capable of being hammered into thin sheets. A piece of gold weighing only 20 grams (slightly less than an ounce) can be hammered into a sheet that will cover more than 68 square feet (6 square meters). The sheet will be only one ten-thousandth of an inch (0.00025 centimeters) thick. Gold foil of this thickness is often used to make the lettering on window signs.

Gold is quite soft. It can usually be scratched by a penny. Its melting point is 1,948.57°F (1,064.76°C) and its boiling point is about 4,900°F (2,700°C). Its density is 19.3 grams per cubic centimeter, almost 20 times that of water.

Two other important properties of gold are its reflectivity and lack of electrical resistance. Both heat and light reflect off gold very well. But an electric current passes through gold very easily.

Chemical Properties

Generally speaking, gold is not very reactive. It does not combine with **oxygen** or dissolve in most acids. It does not react with halogens, such as **chlorine** or **bromine**, very easily.

These chemical properties also account for some important uses of gold. Gold coins, for example, do not corrode (rust) or tarnish very easily. Neither does jewelry or artwork made of gold.

Occurrence in Nature

Gold occurs in nature in both its native state and in compounds. The native state of an element is its free state. It is not combined with any

Pound for Pound: Measuring Gold

Which weighs more: A pound of feathers or a pound of gold? Teachers sometimes try to fool students with this old question. The answer would seem to be easy: a pound is a pound. A pound of feathers and a pound of gold should weigh the same amount.

In terms of weight, gold is weighed differently than most substances, which are measured using the avoirdupois (pronounced a-ver-de-POIZ) system. In the avoirdupois system, there are 16 ounces to the pound. But gold is weighed using the troy system. In the troy system, one pound contains only 12 ounces. So, a pound of feathers (avoirdupois system) weighs four ounces more than a pound of gold (troy system). The weight of other precious metals, like silver and platinum, are also measured using the troy system.

Gold is also weighed in carats. A carat is defined as one fifth of a gram, or 200 milligrams.

Gold is seldom used in a pure form. The metal is too soft. It would bend or break if used pure. Instead, it is used in combination with other metals called alloys. An alloy is a mixture of two or more metals. The mixture has properties different from those of the individual metals.

The amount of gold in an alloy is expressed in carats. Pure gold metal (mixed with no other metal) is said to be 24-carat gold. An alloy that contains 20 parts of gold and 4 parts of silver is 20-carat gold. The "20-carat" designation means the alloy contains 20 parts of gold and 4 parts of something else (silver, in this case).

Gold stored in a national bank can be 24-carat gold. It is never used for any practical purpose. But gold used for any real application is almost always less than 24 carats. It must include other metals that make it stronger and tougher.

other element. The most common compounds of gold are the tellurides. A telluride is a compound of the element **tellurium** and one or more other elements. For example, the mineral calavarite is mostly gold telluride ($AuTe_2$).

At one time, gold was found in chunks or nuggets large enough to see. People mined gold by picking it out of streams and rivers. In fact, gold was once very common in some parts of the world. People valued it not because it was rare, but because it was so beautiful.

The abundance of gold in Earth's crust is estimated to be about 0.005 parts per million. That makes it one of the 10 rarest elements in Earth's crust. Gold is thought to be much more common in the oceans. Some people believe as much as 70 million tons of gold are dissolved in seawater. They also think there may be another 10 billion tons on the bottom of the oceans. So far, however, no one has found a way to mine this gold.

As of 2008, the leading producers of the metal included China, South Africa, the United States, Australia, Peru, and Russia. According to the U.S. Geological Survey (USGS), the price of gold rose 29 percent between 2007 and 2008. In the United States, gold is mined primarily in western states, including Nevada, California, and Alaska.

Isotopes

There is only one naturally occurring isotope of gold: gold-197. Isotopes are two or more forms of an element. Isotopes differ from each other according to their mass number. The number written to the right of the element's name is the mass number. The mass number represents the number of protons plus neutrons in the nucleus of an atom of the element. The number of protons determines the element, but the number of neutrons in the atom of any one element can vary. Each variation is an isotope.

Forty-nine radioactive isotopes of gold are known also. A radioactive isotope is one that breaks apart and gives off some form of radiation. Radioactive isotopes are produced when very small particles are fired at atoms. These particles stick in the atoms and make them radioactive.

One radioactive isotope of gold is widely used in medicine, gold-198. This isotope has two major uses. First, it can be used to study the liver. It is made into a form known as colloidal gold, which consists of very fine particles of gold mixed in a liquid solution. The colloidal gold is injected into the patient's body, where it travels to the liver. There, it can be detected because of the radiation it gives off. The radiation can be used to tell if the liver is functioning normally or not.

Colloidal gold is also used to treat medical problems. In some forms of cancer, the body develops large amounts of liquid in the space around the stomach and intestines (the peritoneum). Some doctors treat this collection of liquid by injecting colloidal gold into the peritoneum. It is not able to leave the peritoneum and go into the stomach and intestines. While in the peritoneum, the colloidal gold gives off radiation. The radiation kills cancer cells that cause the accumulation of fluid.

Extraction

There are at least two main ways to remove gold from its ores. One is to mix an ore with **mercury** metal. Mercury combines with gold in the ore to form an amalgam. An amalgam is a mixture of two or more metals, one of which is mercury. The gold amalgam is then removed from the ore. It is heated to drive off the mercury. Pure gold remains.

Gold ores can also be treated with potassium cyanide (KCN) or some other kind of cyanide. The gold combines with the cyanide to form a new compound, gold cyanate. The gold cyanate is then treated with an active metal, such as zinc. The active metal replaces gold in the compound, leaving pure gold.

Uses

The World Gold Council has estimated that about 140,000 short tons (about 125,000 metric tons) of gold has been mined throughout history. Of that amount, about 10 percent has been used for industrial, research, health, and other "dissipative" uses. Dissipative means that the gold was gone once it was used. It was made into devices that were eventually thrown away. The gold could not or was not recovered from the devices.

About a third of the gold has been made into gold bars held by national banks. The gold bars are used as security for national money systems. In the United States, for example, the nation's supply of gold is stored at Fort Knox in Kentucky.

Finally, the remaining gold is owned by private individuals. This gold exists in the form of jewelry, art, coins, or bullion. Gold bullion are bars or other large pieces of pure gold.

One of the most famous items using gold is the Olympic gold medal. Athletes from around the world dream of coming in first place at the Olympics. That means they can step up to the winner's podium and wear their gold medal proudly. But the gold medal isn't solid gold. It's actually made out of silver. A thin layer of gold covers the silver. The last time a solid gold medal was used in the Olympics was 1912. For the 2010 Olympic Games in Vancouver, British Columbia, Canada, 2.05 kilograms of gold, 1,950 kilograms of silver, and 903 kilograms of copper will be used in the medals.

Jewelry and art objects account for the largest single use of gold. In 2008, 80 percent of the gold used in the United States was for this purpose. The second largest use of gold (about 8 percent) was for electrical and electronic products, such as electrical contacts and switches, printed circuits, and instruments on space vehicles. Medical and dental applications accounted for most of the remaining gold used.

Olympic champion Michael Phelps holds up one of the eight gold medals he won at the Beijing Olympics in 2008.
AP IMAGES.

Compounds

Few gold compounds have any important commercial uses.

Health Effects

Gold is not required to maintain good health in plants or animals. It can be injected into a plant or animal without causing harmful effects. Some medical and commercial uses are based on this property.

Hafnium

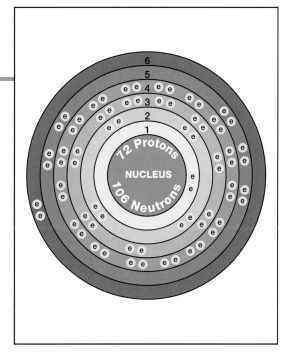

Key Facts

Symbol: Hf

Atomic Number: 72

Atomic Mass: 178.49

Family: Group 4 (IVB); transition metal

Pronunciation: HAF-nee-um

Overview

Hafnium is an element that chemists knew existed, but could not find. They knew it must exist because of an empty space in the periodic table. The periodic table is a chart that shows how chemical elements are related to each other.

By the 20th century, nearly all the spaces in the periodic table had been filled. One of the empty spaces was element 72. A number of chemists searched for the element. Some even claimed they had found it. But these claims could not be confirmed. In fact, it was not until 1923 that element number 72 was finally discovered.

Hafnium is a shiny, silvery-white metal. It is always found with another chemical element, **zirconium**. The two elements are very much alike. In fact, their similarity is the reason that it took so long to find hafnium.

Hafnium has only a few applications. Probably its most important use is in nuclear power plants. A nuclear power plant is a facility where energy released from nuclear fission reactions is used to generate electricity.

WORDS TO KNOW

Alloy: A mixture of two or more metals with properties different from those of the individual metals.

Ductile: Capable of being drawn into thin wires.

Half life: The time it takes for half of a sample of a radioactive element to break down.

Isotopes: Two or more forms of an element that differ from each other according to their mass number.

Periodic table: A chart that shows how the chemical elements are related to each other.

Radioactive isotope: An isotope that breaks apart and gives off some form of radiation.

Discovery and Naming

In the early 1900s, scientists had found a new way to identify elements. This method is called X-ray diffraction analysis. Here is how this method works:

A stream of electrons is fired at a metal plate. The electrons cause the metal plate to give off X rays. The process is similar to what happens when a person has a chest X ray or an X ray of a broken bone.

The kind of X ray produced depends on the metal used. Each metal produces its own special X-ray pattern. In fact, the pattern produced can be used to identify a metal.

In 1923 Dutch physicist Dirk Coster (1889–1950) and Hungarian chemist George Charles de Hevesy (1889–1966) found element 72 by X-ray analysis. The element was present in a piece of Norwegian zircon. Zircon also contains the mineral zirconium.

Chemists later developed a better understanding about the relationship of zirconium and hafnium. These two elements are as alike as any two elements in the periodic table. They have nearly identical chemical and physical properties. This similarity explains why it took so long to find hafnium. Chemists had probably discovered hafnium before 1923, but thought it was zirconium. The differences in the X-ray patterns of the two elements finally proved that hafnium was different from zirconium.

Physical Properties

Hafnium is a bright, silvery-gray metal that is very ductile. Ductile means capable of being drawn into thin wires. Its melting point is about 3,900°F

(2,150°C) and its boiling point about 9,700°F (5,400°C). Its density is 13.1 grams per cubic centimeter, about 13 times that of water.

The physical property of greatest interest for hafnium is how it responds to neutrons. A neutron is a very small particle found in the nucleus (center) of an atom. Neutrons are used to make nuclear fission reactions occur. Nuclear fission reactions take place when a neutron strikes a large atom, such as an atom of **uranium**. The neutron makes the atom break apart. In the process, a large amount of energy is released. That energy can be converted to electricity.

In order to make electricity from nuclear fission, the fission reaction must be carefully controlled. To do that, the number of neutrons must also be kept under close control. Hafnium has the ability to absorb ("soak up") neutrons very easily. It is used in rods that control how fast a fission reaction takes place. This property is one of the few ways in which hafnium differs from zirconium. Although hafnium is very good at

Hafnium is used in nuclear power plants. IMAGE COPYRIGHT 2007, KRISTIN SMITH. USED UNDER LICENSE FROM SHUTTERSTOCK.COM.

absorbing neutrons, zirconium hardly absorbs neutrons at all. Neutrons pass right through it.

Chemical Properties

Like zirconium, hafnium is not very reactive. It does not combine easily with **oxygen** in the air or react with water or cold acids. It may be more active with hot acids, however.

Occurrence in Nature

Hafnium is a moderately common element in Earth's crust. Its abundance is estimated to be about 5 parts per million. That makes it about as abundant as **bromine**, uranium, or **tin**.

Hafnium is always found with ores of zirconium in the earth. The most common of those ores are zircon and baddeleyite.

Isotopes

Hafnium has six naturally occurring isotopes: hafnium-174, hafnium-176, hafnium-177, hafnium-178, hafnium-179, and hafnium-180. The first of these isotopes is radioactive. It has a half life of an astounding two quadrillion years. (That's 2 followed by 15 zeroes!) Isotopes are two or more forms of an element. Isotopes differ from each other according to their mass number. The number written to the right of the element's name is the mass number. The mass number represents the number of protons plus neutrons in the nucleus of an atom of the element. The number of protons determines the element, but the number of neutrons in the atom of any one element can vary. Each variation is an isotope.

Artificial radioactive isotopes can also be produced by firing very small particles at atoms. These particles stick in the atoms and make them radioactive. Thirty-six artificial radioactive isotopes of hafnium are known. There are no important commercial applications for any of these isotopes.

Extraction

The greatest problem in working with hafnium is finding a way to separate it from zirconium. Today, chemists know that compounds of hafnium dissolve more easily in some liquids than do compounds of zirconium. This method can be used to separate compounds of the two elements from each other.

Uses and Compounds

Nuclear power plant applications account for the largest use of hafnium metal. A growing use of hafnium metal is in the production of nickel-based superalloys. A superalloy is a mixture of two or more metals with properties superior to those of traditional alloys. They are stronger, more resistant to corrosion (rusting), and able to retain their properties at high temperature without oxidizing (reacting with oxygen in the air). One of their most common uses is in turbine blades used in electrical generating plants and in jet engines.

Hafnium is also used to make binary compounds with interesting properties. A binary compound consists of two elements. These compounds are among the best refractory materials known. A refractory material is one that can withstand very high temperatures. It reflects heat away from itself. Refractory materials are used to line the inside of high-temperature ovens. For example, some alloys are made at temperatures of thousands of degrees in refractory ovens. Some hafnium compounds used to line these furnaces are hafnium boride (HfB_2), hafnium nitride (HfN), and hafnium oxide (HfO_2).

Health Effects

Both hafnium and its compounds are toxic. They are most dangerous when inhaled. Powdered hafnium metal is also dangerous. It can ignite and explode very easily.

Helium

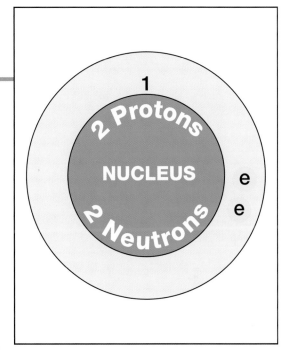

Key Facts

Symbol: He

Atomic Number: 2

Atomic Mass: 4.002602

Family: Group 18 (VIIIA); noble gas

Pronunciation: HEE-lee-um

Overview

Helium is a member of the noble gas family. The noble gases are the elements in Group 18 (VIIIA) of the periodic table. The periodic table is a chart that shows how the elements are related to one another. The noble gases are also called the inert gases. Inert means that an element is not very active. It will not combine with other elements or compounds. In fact, no compounds of helium have ever been made.

Helium is the second most abundant element in the universe. Only **hydrogen** occurs more often than helium. Helium is also the second simplest of the chemical elements. Its atoms consist of two protons, two neutrons, and two electrons. Only the hydrogen atom is simpler than a helium atom. The hydrogen atom has one proton, one electron, and no neutrons.

Helium was first discovered not on Earth, but in the sun. In 1868 French astronomer Pierre Janssen (1824–1907) studied light from the sun during a solar eclipse. He found proof that a new element existed in the sun. He called the element helium.

Helium has some interesting and unusual physical properties. For example, at very low temperatures it can become a superfluid—a material

WORDS TO KNOW

Alpha particles: Helium atoms without their electrons.

Inert: Not very active.

Isotopes: Two or more forms of an element that differ from each other according to their mass number.

Liquid air: Air that has been cooled to a very low temperature.

Noble gas: An element in Group 18 (VIIIA) of the periodic table.

Periodic table: A chart that shows how the chemical elements are related to each other.

Radioactive isotope: An isotope that breaks apart and gives off some form of radiation.

Spectroscope: A device for studying the light produced by a heated object.

Superconductor: A material that has no resistance to the flow of electricity; once an electrical current begins flowing in the material, it continues to flow forever.

that behaves very strangely. It can flow upward out of a container, against the force of gravity. It can also squeeze through very small holes that should be able to keep it out. The Nobel Prize in physics for 1996 was awarded to three Americans who discovered superfluidity. They were David M. Lee (1931–), Douglas D. Osheroff (1945–), and Robert C. Richardson (1937–).

For an inactive gas, helium has a surprising number of applications. It is used in low-temperature research, for filling balloons and dirigibles (blimps), to pressurize rocket fuels, in welding operations, in lead detection systems, in **neon** signs, and to protect objects from reacting with **oxygen**.

Discovery and Naming

One of the most powerful instruments for studying chemical elements is the spectroscope. The spectroscope is a device for studying the light produced by a heated object. For example, a lump of **sodium** metal will burn with a yellow flame. The flame looks quite different, however, when viewed through a spectroscope.

A spectroscope contains a triangular piece of glass (called a prism) that breaks light into its basic parts. These basic parts consist of a series of colored lines. In the case of sodium, the yellow light is broken into a series of yellow lines. These lines are called the element's *spectrum*. Every element has its own distinctive spectrum.

The spectroscope gives scientists a new way of studying elements. They can identify an element by recognizing its distinctive spectral lines even when they can't actually see the element itself.

An Element on the Sun This principle led to the discovery of helium. In 1868, Pierre Janssen visited India in order to observe a full eclipse of the sun. A solar eclipse occurs when the Moon comes between the sun and Earth. The Moon blocks nearly all of the sun's light. All that remains is a thin outer circle (corona) of sunlight around the dark Moon. Solar eclipses provide scientists with an unusual chance to study the sun.

Janssen examined light from the sun with a spectroscope. As he looked at the spectral lines, he was surprised to see some lines that could not be traced to any known element. He concluded that there must be an element on the sun that had never been seen on Earth. The name helium was later suggested for this element. The name comes from the Greek word *helios* for "sun."

Chemists did not know what to make of Janssen's discovery. Was there an element on the sun that did not exist on Earth? Had he made a mistake? Some scientists even made fun of Janssen.

Helium on Earth For the next 30 years, chemists looked for helium on Earth. Then, in 1895, the English physicist Sir William Ramsay (1852–1916) found helium in a mineral of the element **uranium**. Credit for the earthly discovery of helium is sometimes given to two other scientists also. Swedish chemists Per Teodor Cleve (1840–1905) and Nils Abraham Langlet also discovered helium at about the same time in a mineral called cleveite.

Ramsay did not know why helium occurred in an ore of uranium. Some years later, the reason for that connection became obvious. Uranium is a radioactive element—an element that breaks apart spontaneously. It releases radiation and changes into a new element.

Ernest Rutherford's Work One form of radiation produced by uranium consists of alpha particles. Alpha particles are tiny particles moving at very high rates of speed. In 1907, English physicist Ernest Rutherford (1871–1937) showed that an alpha particle is nothing more than a helium atom without its electrons. As uranium atoms broke apart, then, they

Ernest Rutherford. LIBRARY OF CONGRESS.

gave off alpha particles (helium atoms). That is the reason helium was first found on Earth in connection with uranium ores.

Rutherford was one of the great scientific figures of the 20th century. He made a number of important discoveries about the structure of atoms and about radioactivity. For example, he found that an atom consists of two distinct parts, the nucleus and the electrons. He also discovered one form of radiation given off by radioactive materials: alpha particles. Alpha particles, he found, are simply helium atoms without their electrons.

Rutherford was also the first scientist to change one element into another. He accomplished this by bombarding **nitrogen** gas with alpha particles. Rutherford found that oxygen was formed in this experiment. He had discovered a way to convert one element (nitrogen) into a different element (oxygen). The method Rutherford used later became a standard procedure used by many other scientists.

Physical Properties

Helium is a colorless, odorless, tasteless gas. It has a number of unusual properties. For example, it has the lowest boiling point of any element, –452.0°F (–268.9°C). The boiling point for a gas is the temperature at which the gas changes to a liquid. The freezing point of helium is –458.0°F (–272.2°C). Helium is the only gas that cannot be made into a solid simply by lowering the temperature. It is also necessary to increase the pressure on the gas in order to make it a solid.

At a temperature of about –456°F (–271°C), helium undergoes an unusual change. It remains a liquid, but a liquid with strange properties. Superfluidity is one of these properties. The forms of helium are so different that they are given different names. Above –456°F (–271°C), liquid helium is called helium I; below that temperature, it is called helium II.

Chemical Elements, 2nd Edition

Chemical Properties

Helium is completely inert. It does not form compounds or react with any other element.

Occurrence in Nature

Helium is the second most abundant element after hydrogen in the universe and in the solar system. About 11.3 percent of all atoms in the universe are helium atoms. By comparison, about 88.6 percent of all atoms in the universe are hydrogen. Thus, at least 99.9 percent of all atoms are either hydrogen or helium atoms.

By contrast, helium is much less abundant on Earth. It is the sixth most abundant gas in the atmosphere after nitrogen, oxygen, **argon**, carbon dioxide, and neon. It makes up about 0.000524 percent of the air.

It is probably impossible to estimate the amount of helium in Earth's crust. The gas is produced when uranium and other radioactive elements break down. But it tends to escape into the atmosphere almost immediately.

Isotopes

Two isotopes of helium occur naturally: helium-3 and helium-4. Isotopes are two or more forms of an element. Isotopes differ from each other according to their mass number. The number written to the right of the element's name is the mass number. The mass number represents the number of protons plus neutrons in the nucleus of an atom of the element. The number of protons determines the element, but the number of neutrons in the atom of any one element can vary. Each variation is an isotope.

Six radioactive isotopes of helium have been made also. A radioactive isotope is one that breaks apart and gives off some form of radiation. Radioactive isotopes are produced when very small particles are fired at atoms. These particles stick in the atoms and make them radioactive.

None of the radioactive isotopes of helium has any commercial application.

Extraction

In theory, helium could be collected from liquid air. Liquid air is air that has been cooled to a very low temperature. All of the gases in air have

Party balloons are often filled with helium. IMAGE COPYRIGHT 2009, HFNG. USED UNDER LICENSE FROM SHUTTERSTOCK. COM.

liquefied in liquid air. If the liquid air were allowed to evaporate, the last gas remaining after all other gases had evaporated would be helium. There is too little helium in air to make this process worthwhile, however.

There is a much better source of helium. The gas often occurs along with natural gas in reservoirs deep beneath Earth's surface. When wells are dug to collect the natural gas, helium comes to the surface with the natural gas. Then, helium can be separated from natural gas very easily. The temperature of the mixture is lowered, and the natural gas liquefies and is taken away. Gaseous helium is left behind.

An estimated 78 percent of the world's helium was produced in the United States in 2008. Other producers included Algeria, Qatar, Russia, and Poland. Nineteen U.S. plants extracted helium from natural gas. Those plants were located in Colorado, Kansas, New Mexico, Oklahoma, Texas, Utah, and Wyoming. According to the U.S. Geological Survey (USGS), between 2009 and 2015, about nine new helium plant projects are slated to begin work worldwide. Such plants will be located in the United States (2), Algeria, Australia, China, India, Indonesia, Qatar, and Russia.

For many years, the U.S. government operated the Federal Helium Program. This program was responsible for collecting and storing helium for government use. The main customers for this helium were the Department of Defense (DOD), the National Aeronautics and Space Administration (NASA), and the Department of Energy (DOE). The helium was stored underground in huge natural caves.

In 1996, the government decided to end this program. Helium was no longer regarded as essential to national security. The Bureau of Mines began to sell off the federal reserves.

Uses

One of the most important single uses for helium in the United States is in pressure and purge systems. In many industrial operations, it is necessary to pressurize a system. The easiest way to do that is to pump a gas into the system. But the gas should not be one that will react with other substances in the system. Being inert, helium is a perfect choice. Helium is also used for purging, a process that sweeps away all gas in a container. Again, helium is used because it does not react with anything in the container.

Another major use of helium is in low-temperature cooling systems. This application is based on the fact that liquid helium—at –454°F (–270°C)—is cold enough to cool *anything* else. For example, it is used in superconducting devices.

A superconducting material is one that has no resistance to the flow of electricity. Once an electric current begins to flow in the material, it will continue to flow forever. No energy is wasted in moving electricity from one place to another. Superconducting materials may revolutionize electrical systems worldwide someday. The problem is that superconductivity occurs only at very low temperatures. One way to achieve those temperatures is with liquid helium.

Because of its inactivity, helium is also used in welding systems. Welding is the process by which two metals are heated to high temperatures in order to join them to each other. Welding rarely works well in "normal" air. At high temperatures, the metals may react with oxygen to form metal oxides. If they do, they are less likely to join to each other. If the welding is done in a container of helium, this is not a problem. The metals will not react with helium. They will simply join to each other.

Helium is also used in leak-detection systems. If a leak is suspected in a long pipe, helium can be used to look for that leak. It is pumped into one end of the pipe. A detector is held outside the pipe. The detector is designed to measure whether helium is escaping from the system. The detector is moved along the length of the pipe. It is possible to find out whether there is a leak, where it is, and the extent of the leak. Helium is a good gas to use for this purpose because it does not react with anything in the pipe.

Helium is also used to inflate balloons and other lighter-than-air craft, such as dirigibles (blimps). Helium does not have the lifting power of hydrogen. However, hydrogen is flammable and helium is not.

At one time, people thought that dirigibles would be a popular form of transportation. But that never happened. Blimps are still used for limited purposes, such as advertising at major sports and recreational events.

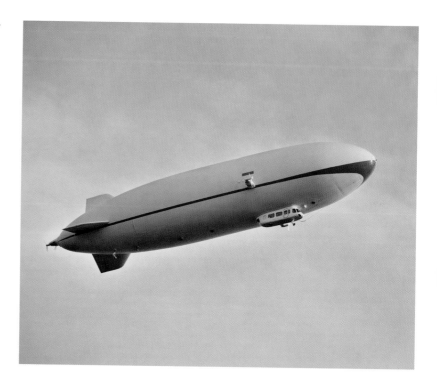

Compounds

No compounds of helium have ever been made.

Health Effects

There are no known health hazards resulting from exposure to helium.

Holmium

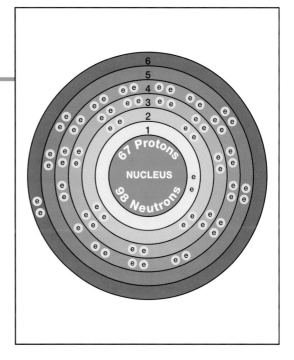

Overview

Holmium occurs in Row 6 of the periodic table. The periodic table is a chart that shows how chemical elements are related to each other. Elements with atomic numbers 57 through 71 are known as the lanthanoids. The name comes from the first element in the series, **lanthanum**. The lanthanoids are also known as rare earth elements. Although lanthanoids are not especially rare, they were once very difficult to separate from each other.

Holmium was discovered by Swedish chemist Per Teodor Cleve (1840–1905) in 1879. He named the element after his birthplace, Stockholm, Sweden. Holmium occurs with other rare earth elements in minerals such as monazite and gadolinite. It can now be separated from other rare earth elements somewhat easily. But no major uses have been found for it or its compounds.

Discovery and Naming

In 1787, a lieutenant in the Swedish army named Carl Axel Arrhenius (1757–1824) was exploring a mine near Ytterby, Sweden. Arrhenius was a "rock hound," a person interested in the study of rocks and minerals. In his explorations, Arrhenius found a rock he had never seen before. He

Key Facts

Symbol: Ho

Atomic Number: 67

Atomic Mass: 164.93032

Family: Lanthanoid (rare earth metal)

Pronunciation: HOL-me-um

WORDS TO KNOW

Ductile: Capable of being drawn into thin wires.

Earth: In mineralogy, a naturally occurring form of an element, often an oxide of the element.

Isotopes: Two or more forms of an element that differ from each other according to their mass number.

Lanthanoids (rare earth elements): The elements in the periodic table with atomic numbers between 57 and 71.

Laser: A device for producing very bright light of a single color.

Malleable: Capable of being hammered into thin sheets.

Periodic table: A chart that shows how the chemical elements are related to each other.

Radioactive isotope: An isotope whose presence in a system can easily be detected.

asked his friend Johan Gadolin (1760–1852), professor of chemistry at the University of Åbo in Finland, to study it. Gadolin discovered in the rock a new mineral, which was given the name ytterite.

Ytterite proved to be a fascinating puzzle for chemists. The mineral contained a number of different "earths." In chemistry, the term earth refers to a naturally occurring compound of an element. For example, magnesia is a naturally occurring compound—an earth—of the element **magnesium.**

Chemists found the earths in ytterite all had very similar properties. For that reason, they had trouble separating them from each other. In fact, it took more than a century to analyze ytterite completely.

In 1879, Per Teodor Cleve was studying an earth taken from yttria called erbia. Erbia had been regarded as a new element for some time. But Cleve separated erbia into three different parts. He called them erbia, holmia, and thulia. Holmia is the earth from which the element holmium comes. For his work, Cleve is given credit for the discovery of holmium.

In Cleve's time, chemical equipment was not very advanced. Chemists usually could not prepare very pure samples of materials. Ten years after the "discovery" of holmium, chemists realized it was actually holmium mixed with another new element, **dysprosium.**

Physical Properties

Like other rare earth elements, holmium is a silvery metal that is soft, ductile, and malleable. Ductile means capable of being drawn into thin wires. Malleable means capable of being hammered into thin sheets. Both properties are common for metals. Holmium also has some rather unusual magnetic and electrical properties.

Holmium has a melting point of 2,680°F (1,470°C) and a boiling point of 4,930°F (2,720°C). Its density is 8.803 grams per cubic centimeter, almost nine times the density of water.

Chemical Properties

Holmium metal tends to be stable at room temperature. In moist air and at higher temperatures, it becomes more reactive. For example, it combines with **oxygen** to form holmium oxide (Ho_2O_3), a yellow solid. Like most other metals, the element also dissolves in acids.

Occurrence in Nature

The abundance of holmium in Earth's crust is estimated to be about 0.7 to 1.2 parts per million. It is less common than most other rare earth elements, but more common than **iodine**, **silver**, **mercury**, and **gold**. The most common ores of holmium are monazite and gadolinite.

Isotopes

Only one naturally occurring isotope of holmium exists: holmium-165. Isotopes are two or more forms of an element. Isotopes differ from each other according to their mass number. The number written to the right of the element's name is the mass number. The mass number represents the number of protons plus neutrons in the nucleus of an atom of the element. The number of protons determines the element, but the number of neutrons in the atom of any one element can vary. Each variation is an isotope.

Fifty radioactive isotopes of holmium are known also. A radioactive isotope is one that breaks apart and gives off some form of radiation. Radioactive isotopes are produced when very small particles are fired at atoms. These particles stick in the atoms and make them radioactive.

None of the radioactive isotopes of holmium has any practical uses.

Extraction

Pure holmium is obtained by reacting **calcium** metal with holmium fluoride (HoF_3):

$$3Ca + 2HoF_3 \rightarrow 3CaF_2 + 2Ho$$

Pure holmium in significant amounts has been available only very recently.

Uses

In the past there were almost no practical uses for holmium and its compounds. However, holmium is now used in specialized lasers. A laser is a device for producing very bright light of a single color. The kind of light produced in a laser depends on the elements of which it is made. Holmium lasers are used to reduce abnormal eye pressure, to treat glaucoma (an eye disorder), and to repair failed glaucoma surgeries.

Another potential use for holmium is a result of its very unusual and strong magnetic properties. It has been used in alloys with other metals to produce some of the strongest magnetic fields ever produced. Holmium also has some limited use in the manufacture of control rods for nuclear power plants. Control rods limit the number of neutrons available to cause the fission of uranium in nuclear reactors, thus controlling the amount of energy produced in the plant.

Compounds

Few holmium compounds have any important commercial uses. The one exception is holmium oxide (Ho_2O_3), used to add a yellowish color to glass, in the manufacture of refractory materials, and as a catalyst for some chemical reactions. A refractory material is a material that can withstand very high temperatures and reflect heat back away from itself. A catalyst is a substance used to speed up a chemical reaction without undergoing any change itself. In 2007, the price of holmium oxide was approximately $750 per kilogram.

Health Effects

Very little is known about the health effects of holmium. It does not appear to pose any serious health hazards to humans or other animals.

Hydrogen

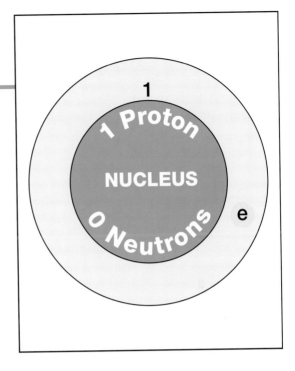

Overview

Hydrogen is the most abundant element in the universe. Nearly nine out of every 10 atoms in the universe are hydrogen atoms. Hydrogen is also common on Earth. It is the third most abundant element after **oxygen** and **silicon**. About 15 percent of all the atoms found on Earth are hydrogen atoms.

Hydrogen is also the simplest of all elements. Its atoms consist (usually) of one proton and one electron.

Hydrogen was first discovered in 1766 by English chemist and physicist Henry Cavendish (1731–1810). Cavendish was also the first person to prove that water is a compound of hydrogen and oxygen.

Some experts believe that hydrogen forms more compounds than any other element. These compounds include water, sucrose (table sugar), alcohols, vinegar (acetic acid), household lye (sodium hydroxide), drugs, fibers, dyes, plastics, and fuels.

Discovery and Naming

Hydrogen was probably "discovered" many times. Many early chemists reported finding a "flammable gas" in some of their experiments. In

Key Facts

Symbol: H

Atomic Number: 1

Atomic Mass: 1.00794

Family: Group 1 (IA)

Pronunciation: HY-dru-jin

WORDS TO KNOW

Catalyst: A substance used to speed up a chemical reaction without undergoing any change.

Halogen: One of the elements in Group 17 (VIIA) of the periodic table.

Isotopes: Two or more forms of an element that differ from each other according to their mass number.

Periodic table: A chart that shows how the chemical elements are related to each other.

Radioactive isotope: An isotope that breaks apart and gives off some form of radiation.

Tracer: A radioactive isotope whose presence in a system can easily be detected.

1671, for example, English chemist Robert Boyle (1627–1691) described experiments in which he added **iron** to hydrochloric acid (HCl) and sulfuric acid (H_2SO_4). In both cases, a gas that burned easily with a pale blue flame was produced.

The problem with these early discoveries was that chemists did not understand the nature of gases very well. They had not learned that there are many kinds of gases. They thought that all the "gases" they saw were some form of air with impurities in it.

Cavendish discovered hydrogen in experiments like those that Boyle performed. He added iron metal to different acids and found that a flammable gas was produced. But Cavendish thought the flammable gas came from the iron and not from the acid. Chemists later showed that iron is an element and does not contain hydrogen or anything else. Therefore, the hydrogen in Cavendish's experiment came from the acid:

$$Fe + 2HCl \rightarrow FeCl_2 + H_2$$

Hydrogen was named by French chemist Antoine-Laurent Lavoisier (1743–1794). Lavoisier is sometimes called the father of modern chemistry because of his many contributions to the science. Lavoisier suggested the name hydrogen after the Greek word for "water former" (that which forms water).

Physical Properties

Hydrogen is a colorless, odorless, tasteless gas. Its density is the lowest of any chemical element, 0.08999 grams per liter. By comparison, a liter of air weighs 1.29 grams, 14 times as much as a liter of hydrogen.

Hydrogen changes from a gas to a liquid at a temperature of –422.99°F (–252.77°C) and from a liquid to a solid at a temperature of –434.6°F (–259.2°C). Hydrogen gas is slightly soluble in water, alcohol, and a few other common liquids.

Chemical Properties

Hydrogen burns in air or oxygen to produce water:

$$2H_2 + O_2 \rightarrow 2H_2O$$

It also combines readily with other non-metals, such as **sulfur**, **phosphorus**, and the halogens. The halogens are the elements that make up Group 17 (VIIA) of the periodic table. They include **fluorine**, **chlorine**, **bromine**, **iodine**, and **astatine**. As an example:

$$H_2 + Cl_2 \rightarrow 2HCl$$

Occurrence in Nature

Hydrogen occurs throughout the universe in two forms. First, it occurs in stars. Stars use hydrogen as a fuel with which to produce energy. The process by which stars use hydrogen is known as fusion. Fusion is the process by which two or more small atoms are pushed together to make one larger atom. In most stars, the primary fusion reaction that occurs is:

$$4H \rightarrow 1He$$

This equation shows that four hydrogen atoms are squeezed together (fused) to make one helium atom. In this process, enormous amounts of energy are released in the form of heat and light.

Hydrogen also occurs in the "empty" spaces between stars. At one time, scientists thought that this space was really empty, that it contained no atoms of any kind. But, in fact, this interstellar space (space between stars) contains a small number of atoms, most of which are hydrogen atoms. A cubic mile of interstellar space usually contains no more than a handful of hydrogen and other atoms.

Hydrogen occurs on Earth primarily in the form of water. Every molecule of water (H_2O) contains two hydrogen atoms and one oxygen atom. Hydrogen is also found in many rocks and minerals. Its abundance is estimated to be about 1,500 parts per million. That makes hydrogen the 10th most abundant element in Earth's crust.

Hydrogen also occurs to a very small extent in Earth's atmosphere. Its abundance there is estimated to be about 0.000055 percent.

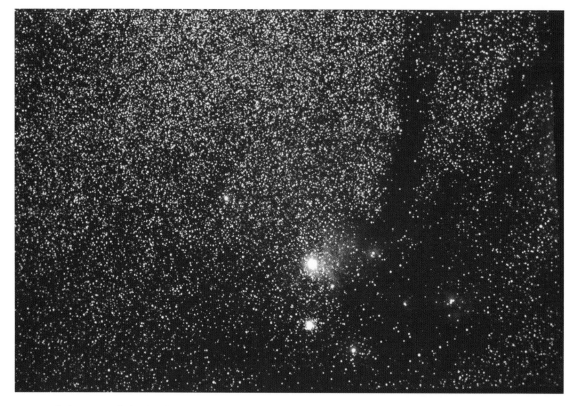

Stars use hydrogen as a fuel with which to produce energy. Antares—the brightest star in the constellation Scorpius—is shown.
© RONALD ROYER/SCIENCE PHOTO LIBRARY, NATIONAL AUDUBON SOCIETY COLLECTION/PHOTO RESEARCHERS, INC.

Hydrogen is not abundant in the atmosphere because it has such a low density. Earth's gravity is not able to hold on to hydrogen atoms very well. Hydrogen floats away into outer space very easily. Most of the hydrogen that was once in the atmosphere has now escaped into outer space.

Isotopes

There are three isotopes of hydrogen: hydrogen-1, hydrogen-2, and hydrogen-3. Isotopes are two or more forms of an element. Isotopes differ from each other according to their mass number. The number written to the right of the element's name is the mass number. The mass number represents the number of protons plus neutrons in the nucleus of an atom of the element. The number of protons determines the element, but the number of neutrons in the atom of any one element can vary. Each variation is an isotope.

The three isotopes of hydrogen have special names.

- Hydrogen-1 is sometimes called protium. It is the simplest and most common form of hydrogen. Protium atoms all contain one proton and one electron. About 99.9844 percent of the hydrogen in nature is protium.

- Hydrogen-2 is known as deuterium. A deuterium atom contains one proton, one electron, and one neutron. About 0.0156 percent of the hydrogen in nature is deuterium.

- The third isotope of hydrogen, hydrogen-3, is tritium. An atom of tritium contains one proton, one electron, and two neutrons. There are only very small traces of tritium in nature.

Tritium is a radioactive isotope. A radioactive isotope is one that breaks apart and gives off some form of radiation. Some radioactive isotopes (such as tritium) occur in nature. They can also be produced in the laboratory. Very small particles are fired at atoms. These particles stick in the atoms and make them radioactive. Tritium is a widely used isotope and is now made in large amounts in the laboratory.

Tritium is widely used as a tracer in both industry and research. A tracer is a radioactive isotope whose presence in a system can be easily detected. The isotope is injected into the system at some point. Inside the system, the isotope gives off radiation. That radiation can be followed by means of detectors placed around the system.

Tritium is popular as a tracer because hydrogen occurs in so many different compounds. For example, suppose a scientist wants to trace the movement of water through soil. The scientist can make up a sample of water made with tritium instead of protium. As that water moves through the soil, its path can be followed by means of the radioactivity the tritium gives off.

Tritium is also used in the manufacture of fusion bombs. A fusion bomb is also known as a hydrogen bomb. In a fusion bomb, small atoms are squeezed together (fused) to make a larger atom. In the process, enormous amounts of energy are given off. For example, the first fusion bomb tested by the United States in 1952 had the explosive power of 15 million tons of TNT. A type of fusion bomb fuses tritium with deuterium to make helium atoms:

$$^{3}\text{H} + {}^{2}\text{H} \rightarrow \text{He}$$

Extraction

The obvious source for hydrogen is water. Earth has enough water to supply people's need for hydrogen. The problem is that it takes a lot of energy to split a water molecule:

$$2H_2O \text{ —electrical current} \rightarrow 2H_2 + O_2$$

In fact, it simply costs too much to make hydrogen by this method. The cost of electricity is too high. So it is not economical to make hydrogen by splitting water.

A number of other methods can be used to produce hydrogen, however. For example, steam can be passed over hot charcoal (nearly pure **carbon**):

$$H_2O + C \text{ —heat} \rightarrow CO + H_2$$

The same reaction can be used with steam and other carbon compounds. For example, using methane, or natural gas (CH_4), the reaction is:

$$H_2O + CH_4 \text{ —heat} \rightarrow CO + 3H_2$$

Hydrogen can also be made by the reaction between carbon monoxide (CO) and steam:

$$CO + H_2O \text{ —heat} \rightarrow CO_2 + H_2$$

Because hydrogen is such an important element, many other methods for producing it have been invented. However, the preceding methods are the least expensive.

Uses

The most important single use of hydrogen is in the manufacture of ammonia (NH_3). Ammonia is made by combining hydrogen and nitrogen at high pressure and temperature in the presence of a catalyst. A catalyst is a substance used to speed up a chemical reaction. The catalyst does not undergo any change during the reaction:

$$N_2 + 3H_2 \text{ —pressure, heat, catalyst} \rightarrow 2NH_3$$

Ammonia is a very important compound. It is used in making many products, the most important of which is fertilizer.

Hydrogen is also used for a number of similar reactions. For example, it can be combined with carbon monoxide to make methanol—methyl alcohol, or wood alcohol (CH_3OH):

$$CO + H_2 \longrightarrow \text{with catalyst} \longrightarrow CH_3OH$$

Like ammonia, methanol has a great many practical uses in a variety of industries. The most important use of methanol is in the manufacture of other chemicals, such as those from which plastics are made. Small amounts are used as additives to gasoline to reduce the amount of pollution released to the environment. Methanol is also used widely as a solvent (to dissolve other materials) in industry.

Another important use of hydrogen is in the production of pure metals. Hydrogen gas is passed over a hot metal oxide to produce the pure metal. For example, **molybdenum** can be prepared by passing hydrogen over hot molybdenum oxide:

$$MoO_2 + 2H_2 \longrightarrow \text{heat} \longrightarrow 2H_2O + Mo$$

Hydrogenation is an important procedure to the food industry. In hydrogenation, hydrogen is chemically added to another substance. The reaction between carbon monoxide and hydrogen is an example of hydrogenation. Liquid oils are often hydrogenated. Hydrogenation changes the liquid oil to a solid fat. Most kitchens contain foods with hydrogenated or partially hydrogenated oils. Vegetable shortening is a good example. Hydrogenation makes it easier to pack and transport oils.

Hydrogen is also used in oxyhydrogen ("oxygen + hydrogen") and atomic hydrogen torches. These torches produce temperatures of a few thousand degrees. At these temperatures, it is possible to cut through steel and most other metals. These torches can also be used to weld (join together with heat) two metals.

Another use for hydrogen is in lighter-than-air balloons. Hydrogen is the least dense of all gases. So a balloon filled with hydrogen can lift very large loads. Such balloons are not used to carry people anymore. The danger of fire or explosion is too great.

The best example of this is the hydrogen fire that destroyed the German airship *Hindenburg,* as it was landing in New Jersey on May 6, 1937 (see sidebar and photo on pages 258 and 259). Today, hydrogen balloons are used for lifting weather instruments into the upper atmosphere.

One of the best known uses of hydrogen is as a rocket fuel. Many rockets obtain the power they need for lift-off by burning hydrogen with pure oxygen in a closed tank. The energy produced by this reaction provides thrust to the rocket.

The Loss of the *Hindenburg*

The *Hindenburg* was Germany's largest passenger airship. It was built in 1936 as a luxury liner, and it made the trip to the United States faster than an ocean liner.

The *Hindenburg* was designed to be filled with helium, a safer gas than the highly flammable hydrogen. But in those pre-World War II days, the United States suspected that Germany's new leader, Adolf Hitler (1889–1945), had military plans for helium-filled ships. So the United States refused to sell helium to the Zeppelin airship company. Seven million cubic feet of hydrogen were used instead. This made the crew very nervous about the potential for fire. Passengers were even checked for matches as they boarded!

On May 3, 1937, the *Hindenburg* left Frankfurt, Germany, for Lakehurst, New Jersey. It traveled over the Netherlands, down the English Channel, through Canada, and into the United States. Bad weather forced the ship to slow down several times, lengthening the trip. But it finally approached the field in Lakehurst around 7 PM on May 6.

After several minutes of maneuvers due to rain and wind, crewmen dropped ropes to the ground at 7:21. The ship was 200 feet (61 meters) above ground. Four minutes later, a small flame emerged on the skin of the ship, and crewmen heard a pop and felt a shudder. Seconds later, the *Hindenburg* exploded. Flaming hydrogen blasted out of the top. Within 32 seconds, the entire airship had burned, the framework had collapsed, and the entire ship lay smoldering on the ground. Thirty-six people died. Amazingly, 62 survived.

Although claims of sabotage have always surrounded the *Hindenburg* tragedy, American and German investigators both agreed it was an accident. Both sides concluded that the airship's hydrogen was ignited probably by some type of atmospheric electric discharge. Witnesses had noticed some of the skin of the ship flapping; they also observed the nose of the ship rise suddenly. Both indicate the likelihood that free hydrogen had escaped. The *Hindenburg* disaster ended lighter-than-air airship travel for many decades.

The Hydrogen Economy Some people believe that hydrogen can help to end the world's dependence on burning fossil fuels (coal, oil, and natural gas) for energy. They have talked about the day when the age of fossil fuels will be replaced by a hydrogen economy.

"The Hydrogen economy" refers to a world in which the burning of hydrogen will be the main source of energy and power. Hydrogen seems like a good choice for helping to meet future energy needs. When it burns, it produces only water:

$$2H_2 + O_2 \rightarrow 2H_2O$$

A lot of energy is produced in this reaction. That energy can be used to operate cars, trucks, trains, boats, and airplanes. It can be used as a source of heat for keeping people warm and running chemical reactions.

So why doesn't a hydrogen economy exist today? The answer is easy. It is still too expensive to make hydrogen gas. No one has found a way to remove hydrogen from water or some other source at a low cost. It is still cheaper to mine for coal or drill for oil than to make hydrogen.

Various automobile companies have researched and created prototypes of hydrogen-powered vehicles. Although a few companies dropped their plans to build such cars in 2009, other corporations announced their intentions to continue such efforts. At that time, most of the hydrogen-powered cars in use in the United States were in California. And more than 60 hydrogen fueling stations were in operation in the country.

Compounds

Millions of hydrogen compounds are known. One of the most important groups of hydrogen compounds is the acids. An acid is any compound that contains hydrogen as its positive part. Common acids include: hydrochloric acid (HCl), sulfuric acid (H_2SO_4), nitric acid (HNO_3), acetic acid ($HC_2H_3O_2$), phosphoric acid (H_3PO_4), and hydrofluoric acid (HF).

Acids are present in thousands of natural substances and artificial products. The following list gives a few examples: vinegar, or acetic acid ($HC_2H_3O_2$); sour milk, or lactic acid ($C_3H_6O_3$); lemons and other citrus fruits, or citric acid ($C_6H_8O_7$); soda water, or carbonic acid (H_2CO_3); battery acid, or sulfuric acid (H_2SO_4); and boric acid (H_3BO_3).

Health Effects

Hydrogen is essential to every plant and animal. Nearly every compound in a living cell contains hydrogen. It is harmless to humans unless taken in very large amounts. In this case, it is dangerous only because it cuts off the supply of oxygen that humans need to breathe.

Indium

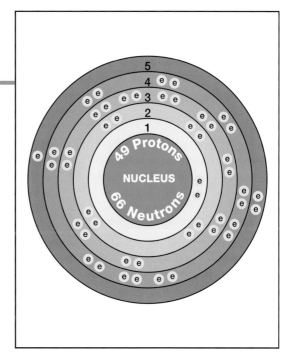

Overview

Indium is part of the aluminum family in Group 13 (IIIA) of the periodic table. The periodic table is a chart that shows how chemical elements are related to each other. Indium was discovered in 1863 by German chemists Ferdinand Reich (1799–1882) and Hieronymus Theodor Richter (1824–1898).

Indium has a number of interesting properties. For example, it has a low melting point for metals, 313.9°F (156.6°C). When pure, it sticks very tightly to itself or to other metals. This property makes it useful as a solder. Solder is a material used to join two metals to each other. Other uses of indium are in the manufacture of batteries, electronic devices, liquid crystal displays (LCD), and in research.

Discovery and Naming

Between 1860 and 1863, indium, **cesium**, **rubidium**, and **thallium** were found using spectroscopy. Spectroscopy is the process of analyzing light produced when an element is heated. The light produced is different for every element. The spectrum (plural: spectra) of an element consists of a series of colored lines. Reich and Richter also produced the first impure sample of indium in 1863.

Key Facts

Symbol: In

Atomic Number: 49

Atomic Mass: 114.818

Family: Group 13 (IIIA); aluminum

Pronunciation: IN-dee-um

WORDS TO KNOW

Absolute zero: The lowest temperature possible, about –459°F (–273°C).

Alloy: A mixture of two or more metals that has properties different from those of the individual metals.

Half life: The time it takes for half of a sample of a radioactive element to break down.

Isotopes: Two or more forms of an element that differ from each other according to their mass number.

Radioactivity: The process by which an isotope or element breaks down and gives off some form of radiation.

Solder: A material used to join two metals to each other.

Spectroscopy: The process of analyzing light produced when an element is heated.

Scientists use a spectroscope in this process. The spectroscope was invented in 1814 by German physicist Joseph von Fraunhofer (1787–1826). Forty years later, German chemists Robert Bunsen (1811–1899) and Gustav Robert Kirchhoff (1824–1887) improved on the instrument. They showed how it could be used to study the chemical elements.

Reich and Richter suggested the name indium for the element they discovered because its main spectral lines are a brilliant indigo blue.

Physical Properties

Indium is a silvery-white, shiny metal with a density of 7.31 grams per cubic centimeter. It is one of the softest metals known, even softer than **lead**. If drawn across a piece of paper, it leaves a mark like a "lead" pencil (which is actually **carbon**).

An unusual property of indium is that it produces a "**tin** cry." A tin cry is a scream-like sound made when the metal is bent.

Indium has a melting point of 313.9°F (156.6°C) and a boiling point of 3,767°F (2,075°C). It has the unusual property of remaining soft and workable at very low temperatures. This property allows it to be used in special equipment needed for temperatures near absolute zero. Absolute zero is the coldest temperature possible. It is about –459°F (–273°C).

Chemical Properties

Indium metal dissolves in acids, but does not react with **oxygen** at room temperature. At higher temperatures, it combines with oxygen to form indium oxide (In_2O_3).

Occurrence in Nature

Indium is relatively rare. Its abundance in Earth's crust is estimated to be about 0.1 part per million. That makes it slightly more abundant than **silver** or **mercury**.

Indium is generally found in ores of **zinc**. The metal is not usually produced in the United States. It is imported mainly from China, Japan, Canada, and Belgium. Use of indium in liquid crystal displays (LCD) products, such as flat-panel monitors, has spiked demand in recent years for indium tin oxide.

Isotopes

Two naturally occurring isotopes of indium exist: indium-113 and indium-115. Isotopes are two or more forms of an element. Isotopes differ from each other according to their mass number. The number written to the right of the element's name is the mass number. The mass number represents the number of protons plus neutrons in the nucleus of an atom of the element. The number of protons determines the element, but the number of neutrons in the atom of any one element can vary. Each variation is an isotope.

Indium-115 is radioactive. A radioactive isotope is one that breaks apart and gives off some form of radiation. Indium-115 has a half life of about 440 trillion years. The half life of a radioactive element is the time it takes for half of a sample of the element to break down. Starting with 10 grams of indium-115 today, only 5 grams would be left 440 trillion years from now.

Seventy artificial radioactive isotopes of indium also exist. Radioactive isotopes are produced when very small particles are fired at atoms. These particles stick in the atoms and make them radioactive. Two of these isotopes are used in medicine. Indium-113 is used to examine the liver, spleen, brain, pulmonary ("breathing") system, and heart and blood system. Indium-111 is used to search for tumors, internal bleeding, abscesses, and infections and to study the gastric (stomach) and blood systems. In both cases, the radioactive isotope is injected into the bloodstream. Inside the body, the isotope gives off radiation. That radiation can be detected by means of a camera or other device. The radiation pattern observed provides information about the organ or system being studied.

Extraction

Indium is obtained in its pure form by separating it from zinc and other elements in zinc ores.

Uses and Compounds

One of the primary uses of indium is in making alloys. An alloy is made by melting and mixing two or more metals. The mixture has properties different from those of the individual metals. Indium has been called a "metal vitamin" in alloys. That means that very small amounts of indium can make big changes in an alloy. For example, very small amounts of indium are sometimes added to **gold** and **platinum** alloys to make them much harder. Such alloys are used in electronic devices and dental materials.

When indium is in its pure form, it sticks very tightly to itself or to other metals, making it useful as a solder. Here, a circuit board is being soldered. IMAGE COPYRIGHT 2009, JOHN STEBBINS. USED UNDER LICENSE FROM SHUTTERSTOCK.COM.

Indium is also added to solders. It reduces the melting point of some solders, strengthens other solders, and prevents still other solders from breaking down too easily.

An important use of indium is in making coatings. For example, some aircraft parts are made of alloys that contain indium. The indium prevents them from wearing out or reacting with oxygen in the air.

Alloys and compounds of indium are also used in making optical (light) devices. For example, indium **gallium** arsenide (InGaAs) is able to convert pulses of light into electrical signals. One application of a device like this is in solar cells. A solar cell is a device used to change sunlight into electrical current. Many scientists think that solar cells may replace coal, oil, and natural gas for many purposes in the future.

Another primary use of indium is in liquid crystal displays (LCDs). Products using LCDs include large flat panel monitors and televisions, cell phones, and digital cameras. Indium is also used in the production of light-emitting diodes (LEDs). A light-emitting diode is a device that gives off light when an electric current passes through it. The numbers on a hand-held calculator are produced with LEDs. In 2004, the U.S. National Park Service replaced all of the conventional lighting at the Thomas Jefferson Memorial in Washington, D.C., with light-emitting diodes containing indium. The conversion reduced the amount of electricity used at the monument by 80 percent and the number of electrical fixtures by 30 percent. At the same time, the total amount of light produced was increased by 20 percent.

Health Effects

Indium is an irritant to the skin, eyes, and respiratory (breathing) system. If taken into the body, it can cause coughing, wheezing, and shortness of breath. Exposure to very high levels of indium may cause damage to the liver and kidneys.

Iodine

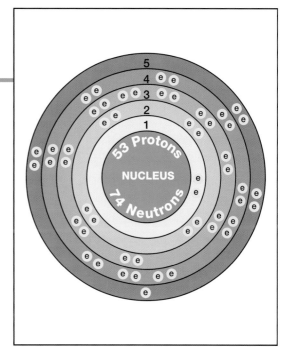

5
4 e e
3
e e   e e   e e
e e   e e
2
e e
1
e e

53 Protons

NUCLEUS

74 Neutrons

Overview

Iodine is the heaviest of the commonly occurring halogens. The halogens are in Group 17 (VIIA) of the periodic table. The periodic table is a chart that shows how chemical elements are related to each other. Iodine's chemical properties are similar to the lighter halogens above it, **fluorine**, **chlorine**, and **bromine**. But its physical appearance is very different. It is a steel-gray solid that changes into beautiful purple vapors when heated.

Iodine was discovered in 1811 by French chemist Bernard Courtois (1777–1838). The element occurs primarily in seawater and in solids formed when seawater evaporates. Its single most important property may be the ability to kill germs. It is used in antiseptics, germicides (products that kill germs), and other medical applications. However, it has a great many other important commercial applications.

Discovery and Naming

One of Bernard Courtois' first jobs was to assist his father in making compounds of **sodium** and **potassium** from seaweed. Seaweed plants take sodium and potassium compounds out of seawater. The compounds become part of the growing seaweed.

Key Facts

Symbol: I

Atomic Number: 53

Atomic Mass: 126.90447

Family: Group 17 (VIIA); halogen

Pronunciation: EYE-uh-dine

WORDS TO KNOW

Antiseptic: A chemical that stops the growth of germs.

Catalyst: A substance used to speed up a chemical reaction without undergoing any change itself.

Halogen: One of the elements in Group 17 (VIIA) of the periodic table.

Isotopes: Two or more forms of an element that differ from each other according to their mass number.

Periodic table: A chart that shows how the chemical elements are related to each other.

Radioactive isotope: An isotope that breaks apart and gives off some form of radiation.

Sublimation: The process by which a solid turns directly to a gas without first melting.

Tincture: A solution made by dissolving a substance in alcohol.

Courtois and his father collected seaweed on the coasts of Normandy and Brittany in France. Then they burned it. Next, they soaked the seaweed ashes in water to dissolve the sodium and potassium compounds. Sulfuric acid was added to react with the unwanted seaweed chemicals. Finally, they allowed the water to evaporate, leaving the compounds behind. These compounds are white crystals, much like ordinary table salt. The compounds were sold to large industrial businesses for use in such products as table salt and baking soda.

One day in 1811, Courtois made a mistake. He added too much sulfuric acid to the mixture. He was amazed to see clouds of beautiful violet vapor rising from the mixture. He decided to study the new material. Eventually, he proved it was a new element. He named the element after its color. In Greek, the word *iodes* means "violet."

Physical Properties

Iodine is one of the most striking and beautiful of all elements. As a solid, it is a heavy, grayish-black, metallic-looking material. When heated, it does not melt. Instead, it sublimes. Sublimation is the process by which a solid turns directly to a gas without first melting. The resulting iodine vapor has a violet color and a harsh odor. If a cold object, such as an **iron** bar, is placed in these vapors, iodine changes back to a solid. It forms attractive, delicate, metallic crystals.

Iodine dissolves only slightly in water. But it dissolves in many other liquids to give distinctive purple solutions. If heated under the proper

A magnified view of a crystal of iodine. © ALFRED PASIEKA/PHOTO RESEARCHERS, INC.

conditions, it can be made to melt at 236.3°F (113.5°C) and to boil at 363°F (184°C). The density of the element is 4.98 grams per cubic centimeter, nearly five times that of water.

Chemical Properties

Like the other halogens, iodine is an active element. However, it is less active than the three halogens above it in the periodic table. Its most common compounds are those of the alkali metals, sodium, and potassium. But it also forms compounds with other elements. It even forms compounds with the other halogens. Some examples are iodine monobromide (IBr), iodine monochloride (ICl), and iodine pentafluoride (IF_5).

Occurrence in Nature

Iodine is not very abundant in Earth's crust. Its abundance is estimated to be about 0.3 to 0.5 parts per million. It ranks in the bottom third of the elements in terms of abundance. It is still more common than **cadmium**, **silver**, **mercury**, and **gold**. Its abundance in seawater is estimated to be even less, about 0.0003 parts per million.

Iodine tends to be concentrated in Earth's crust in only a few places. These places were once covered by oceans. Over millions of years, the oceans evaporated. They left behind the chemical compounds that had been dissolved in them. The dry chemicals left behind were later buried by earth movements. Today, they exist underground as salt mines.

Iodine can also be collected from seawater, brackish water, brine, or sea kelp. Seawater is given different names depending on the amount of solids dissolved in it. Brackish water, for example, has a relatively low percentage of solids dissolved in water. The range that is sometimes given is 0.05 to 3 percent solids in the water. Brine has a higher percentage of dissolved solids. It may contain anywhere from 3 to 20 percent of solids dissolved in water.

Finally, sea kelp is a form of seaweed. As it grows, it takes iodine out of seawater. Over time, sea kelp has a much higher concentration of iodine than seawater. Sea kelp is harvested, dried, and burned to collect iodine. The process is not much different from the one used by Courtois in 1811.

Isotopes

Only one naturally occurring isotope of iodine is known: iodine-127. Isotopes are two or more forms of an element. Isotopes differ from each other according to their mass number. The number written to the right of the element's name is the mass number. The mass number represents the number of protons plus neutrons in the nucleus of an atom of the element. The number of protons determines the element, but the number of neutrons in the atom of any one element can vary. Each variation is an isotope.

Forty-three radioactive isotopes of iodine have been made artificially. A radioactive isotope is one that breaks apart and gives off some form of radiation. Radioactive isotopes are produced when very small particles are fired at atoms. These particles stick in the atoms and make them radioactive.

A number of iodine isotopes are used commercially. In medical applications, these isotopes are injected into the body or given to the patient through the mouth. The isotopes then travel through the body in the bloodstream. As they travel, they give off radiation. That radiation can be detected by using X-ray film. A medical specialist can tell how well the body is functioning by observing the pattern of radiation.

Iodine isotopes are used in many ways. Iodine-123 is used in studies of the brain, kidneys, and thyroid. Iodine-125 is used in studies of the pancreas, blood flow, thyroid, liver, take-up of minerals in bones, and loss of proteins in the body. And iodine-131 is used in studies of the liver, kidneys, blood flow, lungs, brain, pancreas, and thyroid.

The most common iodine isotope used is iodine-131. When iodine (of any kind) enters the body, it tends to go directly to the thyroid. The iodine is then used to make thyroid hormones. If radioactive iodine is used, a doctor can tell how well the thyroid gland is working. If a high level of radiation is given off, the gland may be overactive. If a low level of radiation is given off, the gland may be underactive. In either case, the person may need some treatment to help the thyroid gland work more normally.

Extraction

A number of methods are available for producing iodine. In one method, sodium iodide (NaI) or potassium iodide (KI) from brine wells is treated with chlorine (Cl_2) gas. The more active chlorine replaces iodine from the sodium or potassium iodide, leaving behind free iodine. In another process, silver nitrate ($AgNO_3$) is added to brine. The silver iodide (AgI) that forms in this process precipitates out (settles out of solution) and is treated with metallic iron. The iron replaces silver in the silver iodide, forming iron(III) iodide (FeI_3). The iron(III) iodide is then reacted with chlorine to obtain free iodine. A third method begins with sodium iodate ($NaIO_3$), commonly found in Chilean saltpeter. The sodium iodate is first treated with sodium bisulfite ($NaHSO_3$) to obtain sodium iodide. The sodium iodide is then allowed to react with excess sodium iodate to produce pure iodine.

Uses and Compounds

One of the major uses of iodine and its compounds is in sanitation systems or in making various antiseptics and drugs. Iodine is also used to make

Low Iodine Levels: A Correctable Problem

The amount of iodine in the human body is very small. To calculate the amount of iodine in one's body, one's body weight is divided by 2,500,000. That number is the weight of iodine in the typical human body. For normal people, the amount is about equal to the size of the head of a pin.

That tiny dot of iodine can mean the difference between good and bad health. People who do not have enough iodine may develop serious health problems. At one time, the most common of those problems was a disease known as goiter. Goiter is a large lump in the neck that develops as the thyroid grows out of control. (It can grow as large as a grapefruit.) A goiter develops as the thyroid tries to make certain hormones, but it does not receive enough iodine from the person's diet. So the thyroid keeps expanding, trying to do its job, until it forms a goiter.

A lack of iodine can cause other problems too. For example, thyroid hormones are needed for normal brain development in an unborn child. They are also needed to continue that development after birth. People who do not include enough iodine in their diet do not develop

normally. Today, experts say that low levels of iodine are the leading cause of mental retardation, deafness, mutism (the inability to speak), and paralysis. They also say less serious problems can be blamed on low iodine levels. These problems include lethargy (drowsiness), clumsiness, and learning disabilities.

Low iodine levels can be easily corrected. In most developed countries today, companies that make table salt add a small amount of potassium iodide (KI) to their salt. The salt is labeled "iodized salt." People who use iodized salt get all the iodine they need for normal thyroid function.

But people who live in developing countries may not be able to get iodized salt. The World Health Organization (WHO) is trying to find ways to provide iodine to such people. In 2007, the WHO estimated that 31 percent of people worldwide were not getting enough iodine in their diets. WHO scientists noted that severe iodine deficiency leads many people to develop mental disabilities. The WHO and other organizations are working to ensure that future generations in these regions get the iodine needed to develop and function normally.

dyes, photographic film, and specialized soaps. It is used in some industries as a catalyst. A catalyst is a substance used to speed up a chemical reaction. The catalyst does not undergo any change itself during the reaction.

Health Effects

Iodine can have both favorable and unfavorable effects on living organisms. It tends to kill bacteria and other disease-causing organisms. In fact, this property leads to its use in sanitation systems and as an antiseptic. An antiseptic is a chemical that stops the growth of germs. Not so long ago, tincture of iodine was one of the most popular antiseptics. It was applied to cuts and wounds to prevent infection. Tincture is a solution made by dissolving some substance (such as iodine) in alcohol rather than in water. Today, tincture of iodine has been replaced by other antiseptics.

One reason that tincture of iodine is used less commonly today is that it can also cause problems. In higher doses, iodine can irritate or burn the skin. It can also be quite poisonous if taken internally.

Iodine plays an important role in the health of plants and animals. It is needed to maintain good health and normal growth. In humans, iodine is used to make a group of important compounds known as thyroid hormones. These chemicals are produced in the thyroid gland at the base of the neck. These chemicals control many important bodily functions. A lack of thyroid hormones can result in the disorder known as goiter. Goiter causes a large lump in the neck as the thyroid grows out of control. Iodine is added to table salt today, so goiter is rarely seen in the United States.

Iridium

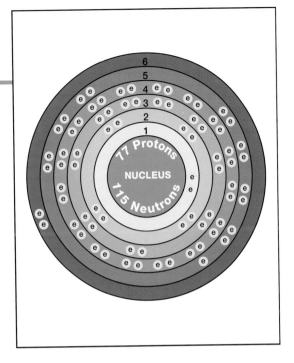

Overview

Iridium is in Group 9 (VIIIB) of the periodic table. The periodic table is a chart that shows how elements are related to one another. Iridium is a transition metal that is also part of the **platinum** family.

The metals in the platinum family are also known as the noble metals. They have this name because they do not react well with other elements and compounds. They appear to be "too superior" to react with most other substances.

In fact, iridium is the most corrosion-resistant metal known. It is not affected by high temperatures, acids, bases, or most other strong chemicals. That property makes it useful in making objects that are exposed to such materials.

Iridium may be a key element in the puzzle of dinosaur extinction. Scientists search for iridium in the soil to track the impact that a giant meteor had on Earth 65 million years ago.

Discovery and Naming

The platinum metals posed a difficult problem for early chemists. These metals often occurred mixed together in the earth. When a scientist

Key Facts

Symbol: Ir

Atomic Number: 77

Atomic Mass: 192.217

Family: Group 9 (VIIIB); transition metal; platinum group

Pronunciation: i-RI-dee-um

WORDS TO KNOW

Alloy: A mixture of two or more metals with properties different from those of any one metal alone.

Aqua regia: A mixture of two strong acids—nitric acid and hydrochloric acid.

Density: The mass of a substance per unit volume.

Ductile: Capable of being drawn into thin wires.

Halogens: Elements in Group 17 (VIIA) of the periodic table.

Isotopes: Two or more forms of an element that differ from each other according to their mass number.

Platinum family (noble metals): A group of elements that occur close to platinum in the periodic table and in Earth's surface.

Radioactive isotope: An isotope that gives off radiation and changes into a new form.

Reactive: Combines with other substances relatively easily.

---

thought that he or she was analyzing a sample of platinum, the sample often contained iridium, **rhodium**, **osmium**, and other metals as well.

The work of French chemist Pierre-François Chabaneau is an example. In the late 1780s, the Spanish government gave its entire supply of platinum to Chabaneau to study. But Chabaneau's experiments puzzled him. Sometimes the platinum he worked with could be hammered into flat plates easily. At other times, it was brittle and shattered when hammered. Chabaneau did not realize that the "platinum" he was studying included various amounts of other noble metals.

In the early 1800s, a number of chemists worked to separate the platinum metals. One of those chemists was an English scientist named Smithson Tennant (1761–1815). Like so many others, Tennant became interested in chemistry at an early age. He is said to have made gunpowder to use in fireworks when he was only nine years old!

In 1803, Tennant attempted to dissolve platinum in aqua regia. Aqua regia is a mixture of two strong acids—nitric acid and hydrochloric acid. He found that most of the platinum metal dissolved, leaving a small amount of black powder. Other chemists had not bothered to study the powder. But Tennant did. He discovered that it had properties very different from those of platinum. He realized he had discovered a new element. He named it iridium, from the Greek goddess Iris, whose symbol is a rainbow. Tennant chose this name because the compounds of iridium have so many different colors. For example, iridium potassium

Small amounts of iridium can be found in meteorites. The Barrington Crater (also known as Meteor Crater), in northern Arizona, was created about 25,000 years ago by a meteorite the size of a large house. It hit the ground at 9 miles per second, and it created a hole 0.7 miles (1.2 kilometers) across and 590 feet (180 meters) deep. IMAGE COPYRIGHT 2009, ALEX NEAUVILLE. USED UNDER LICENSE FROM SHUTTERSTOCK.COM.

chloride (K_2IrCl_6) is dark red, iridium tribromide ($IrBr_3$) is olive-green, and iridium trichloride ($IrCl_3$) is dark green to blue-black.

Physical Properties

Iridium metal is silvery-white with a density of 22.65 grams per cubic centimeter. A cubic centimeter of iridium weighs 22.65 times as much as a cubic centimeter of water. It is one of the two densest elements known (the other being osmium). Iridium has a melting point of 4,429°F (2,443°C) and a boiling point of about 8,130°F (4,500°C). Cold iridium metal cannot be worked easily. It tends to break rather than bend. It becomes more ductile (flexible) when hot. Ductile means capable of being drawn into thin wires. Therefore, it is usually shaped at high temperatures.

Chemical Properties

Iridium is unreactive at room temperatures. When exposed to air, it reacts with oxygen to form a thin layer of iridium dioxide (IrO_2).

$$Ir + O_2 \rightarrow IrO_2$$

At high temperatures, the metal becomes more reactive. Then it reacts with oxygen and halogens to form iridium dioxide and iridium trihalides. For example:

$$2Ir + 3Cl_2 \rightarrow 2IrCl_3$$

Occurrence in Nature

Iridium is one of the rarest elements in Earth's crust. Its abundance is estimated to be about two parts per billion. Interestingly, it is more abundant in other parts of the universe. Iron meteorites, for example, generally contain about 3 parts per million of iridium. Stony meteorites contain less iridium, about 0.64 parts per million.

Iridium usually occurs in combination with one or more other noble metals. Two common examples are osmiridium and iridosmine,

This image from the National Aeronautics and Space Administration (NASA) shows what it might have looked like when an asteroid struck Earth millions of years ago. Scientists theorize that an iridium-rich layer formed on Earth when the asteroid hit. NATIONAL AERONAUTICS AND SPACE ADMINISTRATION (NASA).

combinations of iridium and osmium. The most important sources of iridium metal are Canada, South Africa, Russia, and the state of Alaska.

Iridium's Link to Dinosaur Extinction Iridium occurrences may hold the key to solving the mystery of why the dinosaurs went extinct. The question of what happened to the dinosaurs has long been one of the most interesting and puzzling issues in science. What happened to make these huge reptiles disappear in such a short period of geologic time?

One answer might be found in the Asteroid Disaster Theory. According to this theory, a huge asteroid struck Earth's surface about 65 million years ago. The exploding asteroid threw enormous amounts of dust into the air. The dust blocked out sunlight for more than a year.

As a result, plants on Earth's surface died. Dinosaurs who lived on those plants died out. So did the meat-eating dinosaurs who lived off the plant eaters.

But how is it possible to know if an asteroid really did hit Earth's surface so long ago? Scientists believe they have now found an answer.

In some parts of the earth, scientists have found a layer of crust that contains an unusually high level of iridium metal. Although iridium rarely occurs on Earth, it is found much more commonly in meteors and asteroids. Scientists believe the iridium-rich layer was formed when an asteroid struck Earth's surface.

This "iridium clue" is a key, therefore, to understanding how dinosaurs disappeared from Earth. And could explain why they, and so many other species of plants and animals, died out so suddenly.

Isotopes

Two naturally occurring isotopes of iridium exist: iridium-191 and iridium-193. Isotopes are two or more forms of an element. Isotopes differ from each other according to their mass number. The number written to the right of the element's name is the mass number. The mass number represents the number of protons plus neutrons in the nucleus of an atom of the element. The number of protons determines the element, but the number of neutrons in the atom of any one element can vary. Each variation is an isotope.

Fifty-three radioactive isotopes of iridium have been made. A radioactive isotope is one that gives off radiation and changes into a new form. The only important radioactive isotope of iridium is iridium-192. This isotope has a half-life of about 74 days. A half-life is the time it takes for one half of a sample to break down. Iridium-192 is used to make X-ray photographs of metal castings and to treat cancer.

Extraction

Iridium and the other platinum metals tend to occur together. A series of chemical reactions is used to separate one metal from the other. The other metals are then removed by other techniques. Very little iridium is produced each year, probably no more than a few metric tons. According to the U.S. Geological Survey (USGS), the price of iridium in 2008 was estimated to be approximately $450 per troy ounce.

Uses

One of the primary uses of iridium is in the manufacture of alloys. An alloy is made by melting and mixing two or more metals. An alloy's properties differ from those of the elements that make it up. Iridium is often combined with platinum, for example, to provide a stronger material than the platinum itself. These alloys are very expensive and are used for only special purposes. For instance, the spark plugs used in helicopters are made of a platinum-iridium alloy. Such alloys are also used for electrical contacts, special types of electrical wires, and electrodes.

Iridium metal is also being used as catalysts. Catalysts are substances that speed up a reaction without changing themselves. Iridium catalysts have been used in amazing new products. For example, one kind of iridium catalyst is able to capture sunlight and turn it into chemical energy. That process is similar to the one used by plants in photosynthesis. Finding a synthetic (artificial) way to make photosynthesis happen is one of the great goals of modern chemistry.

Space technology often uses alloys that are too expensive for everyday use. An example is the propulsion systems used for keeping satellites in place. Some of these systems use alloys made of iridium and another platinum metal, **rhenium**. These alloys remain strong at high temperatures and are not attacked by fuels used in the systems.

An interesting use of iridium involves the official world standard for the kilogram. The standard, kept at the International Bureau of Weights and Measures in Paris, France, is a piece of platinum-iridium metal stored in an airtight jar. The standard is made of platinum and iridium to protect it from reacting with oxygen and other chemicals in the air. In this way, the standard's weight will always remain exactly the same.

Compounds

The compounds of iridium have almost no practical applications. A few are used in coloring ceramics because of their striking colors.

Health Effects

Scientists are not aware of any health benefits or risks associated with iridium. Its compounds are thought to be irritating to the skin, eyes, and respiratory system, and toxic if ingested (swallowed).

Iron

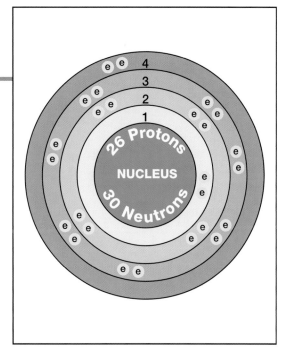

Overview

The period in human history beginning in about 1200 BCE is called the Iron Age. It was at about this time that humans first learned how to use iron metal. But in some ways, one could refer to the current era as the New Iron Age. Iron is probably the most widely used and most important metal today. No other metal is available to replace iron in all its many applications.

Iron is a transition metal. The transition metals are the elements that make up Groups 3 through 12 in the periodic table. The periodic table is a chart that shows how elements are related to one another. The transition metals are typical metals in that they tend to be bright, shiny, silvery solids. They all tend to conduct heat and electricity well. And they usually have high melting points.

Iron normally does not occur as a free element in the earth. In fact, iron was not of much value to humans until they learned how to free iron from its compounds. Once they could do that, humans were able to make tools, weapons, household implements, and other objects out of iron. This step forward marked the beginning of the Iron Age.

Key Facts

Symbol: Fe

Atomic Number: 26

Atomic Mass: 55.845

Family: Group 8 (VIIIB); transition metal

Pronunciation: EYE-urn

WORDS TO KNOW

Alloy: A mixture of two or more metals with properties different from those of the individual metals.

Catalyst: A substance used to speed up a chemical reaction without undergoing any change itself.

Ductile: Capable of being drawn into thin wires.

Isotopes: Two or more forms of an element that differ from each other according to their mass number.

Malleable: Capable of being hammered into thin sheets.

Periodic table: A chart that shows how the chemical elements are related to each other.

Radioactive isotope: An isotope that breaks apart and gives off some form of radiation.

Tensile: Capable of being stretched without breaking.

Tracer: A radioactive isotope whose presence in a system can easily be detected.

Transition metal: An element in Groups 3 through 12 of the periodic table.

Workability: The ability to work with a metal to get it into a desired shape or thickness.

Iron is most valuable not as a pure metal, but in alloys. An alloy is made by melting and mixing two or more metals. The mixture has properties different from those of the individual metals. The best known and most widely used alloy of iron is steel. Steel contains iron and at least one other element. Today, specialized steels of all kinds are available for many different applications.

Discovery and Naming

Ancient Egyptians had learned how to use iron before the First Dynasty, which began in about 3400 BCE. The Egyptians probably found the iron in meteorites. Meteorites are chunks of rock and metal that fall from the sky. Some meteorites are very rich in iron. The Egyptians made tools and jewelry out of iron.

Iron was also known to early Asian civilizations. In Delhi, India, for example, a pillar made out of iron built in 415 CE still stands. It weighs about 7 short tons (6.5 metric tons) and remains in good condition after nearly 1,600 years.

Early Chinese civilizations also knew about iron. Workers learned to produce iron as early as 200 BCE. A number of iron objects, including cannons, remain from the Han period (202 BCE to 221 CE.)

Some meteorites are very rich in iron. Here, children play on the Williamette meteorite in Hayden Planetarium in New York City, in 1939. Named for the Williamette Valley in Oregon where it was discovered, the meteorite weighs 15.5 tons and is the largest one yet found in the United States. © CORBIS-BETTMANN.

The Bible also includes many mentions of iron. For example, a long passage in the book of Job describes the mining of iron. Other passages tell about the processing of iron ore to obtain iron metal.

By the time of the Roman civilization, iron had become an essential metal. The historian Pliny (23–79 CE) described the role of iron in Rome:

> It is by the aid of iron that we construct houses, cleave rocks, and perform so many other useful offices of life. But it is with iron also that wars, murders, and robberies are effected, and this, not only hand to hand, but from a distance even, by the aid of weapons and winged

weapons, now launched from engines, now hurled by the human arm, and now furnished with feathery wings.

Even from the earliest days, humans probably seldom used iron in a pure form. It was difficult to make iron that was free of impurities, such as **carbon** and other elements. More important, however, it became obvious that iron *with* impurities was a stronger metal than iron *without* impurities.

It was not until 1786, however, that scientists learned what it was in steel that made it a more useful metal than iron. Three researchers, Gaspard Monge (1746–1818), C. A. Vandermonde, and Claude Louis Berthollet (1748–1822), solved the puzzle. They found that a small amount of carbon mixed with iron produced a strong alloy. That alloy was steel. Today, the vast amount of iron used in so many applications is used in the form of steel, not pure iron.

The chemical symbol for iron is Fe. That symbol comes from the Latin name for iron, *ferrum.*

Physical Properties

Iron is a silvery-white or grayish metal. It is ductile and malleable. Ductile means capable of being drawn into thin wires. Malleable means capable of being hammered into thin sheets. It is one of only three naturally occurring magnetic elements. The other two are **nickel** and **cobalt**.

Iron has a very high tensile strength. Tensile strength is a measure of a substance's ability to be stretched without breaking. Iron is also very work-able. Workability is the ability to bend, roll, hammer, cut, shape, form, and otherwise work with a metal to get it into a desired shape or thickness.

The melting point of pure iron is 2,797°F (1,536°C) and its boiling point is about 5,400°F (3,000°C). Its density is 7.87 grams per cubic centimeter, almost eight times that of water. The melting point, boiling point, and other physical properties of steel alloys may be quite different from those of pure iron.

Chemical Properties

Iron is a very active metal. It readily combines with **oxygen** in moist air. The product of this reaction, iron oxide (Fe_2O_3), is known as rust. Iron also reacts with very hot water and steam to produce **hydrogen** gas. It also dissolves in most acids and reacts with many other elements.

Chemical Elements, 2nd Edition

Iron is one of only three naturally occurring magnetic elements. Here, a bar magnet with iron filings shows a magnetic field pattern. IMAGE COPYRIGHT 2009, AWE INSPIRING IMAGES. USED UNDER LICENSE FROM SHUTTERSTOCK.COM.

Occurrence in Nature

Iron is the fourth most abundant element in Earth's crust. Its abundance is estimated to be about 5 percent. Most scientists believe that Earth's core consists largely of iron. Iron is also found in the sun, asteroids, and stars.

The most common ores of iron are hematite, or ferric oxide (Fe_2O_3); limonite, or hydrated ferric oxide; magnetite, or iron oxide (Fe_3O_4); and siderite, or iron carbonate ($FeCO_3$). An increasingly important source of iron is taconite. Taconite is a mixture of hematite and silica (sand). It contains about 25 percent iron.

As of 2008, the largest producers of iron ore in the world were Brazil, Australia, India, Russia, and China. According to the U.S. Geological

Survey, 98 percent of the usable iron ore shipped in the United States came from mines in Michigan and Minnesota. The value of those shipments alone was estimated at more than $3 billion.

Isotopes

There are four naturally occurring isotopes of iron: iron-54, iron-56, iron-57, and iron-58. Isotopes are two or more forms of an element. Isotopes differ from each other according to their mass number. The number written to the right of the element's name is the mass number. The mass number represents the number of protons plus neutrons in the nucleus of an atom of the element. The number of protons determines the element, but the number of neutrons in the atom of any one element can vary. Each variation is an isotope.

Twenty-two radioactive isotopes of iron are known also. A radioactive isotope is one that breaks apart and gives off some form of radiation. Radioactive isotopes are produced when very small particles are fired at atoms. These particles stick in the atoms and make them radioactive.

Two radioactive isotopes of iron are used in medical and scientific research. They are iron-55 and iron-59. These isotopes are used primarily as tracers in studies on blood. A tracer is a radioactive isotope whose presence in a system can easily be detected. The isotope is injected into the system. Inside the system, the isotope gives off radiation. That radiation can be followed by detectors placed around the system. Iron-55 and iron-59 are used to study the way in which red blood cells develop in the body. These studies can be used to tell if a person's blood is healthy.

Extraction

Iron goes through a number of stages between ore and final steel product. In the first stage, iron ore is heated with limestone and coke (pure carbon) in a blast furnace. A blast furnace is a very large oven in which the temperature may reach 2,700°F (1,500°C). In the blast furnace, coke removes oxygen from iron ore:

$$3C + 2Fe_2O_3 \xrightarrow{\text{heat}} 3CO_2 + 4Fe$$

The limestone removes impurities in the iron ore.

Iron produced by this method is about 91 to 92 percent pure. The main impurity left is carbon from the coke used in the furnace. This form of iron is known as pig iron. Pig iron is generally too brittle (it breaks too easily) to be used in most products.

A number of methods have been developed for purifying pig iron. A common method used is the basic oxygen process. In this process, pig iron is melted in a large oven. Then pure oxygen gas is blown through the molten pig iron. The oxygen burns off much of the carbon in the pig iron:

$$C + O_2 \text{ —heat} \rightarrow CO_2$$

A small amount of carbon remains in the iron. The iron produced in this reaction is known as steel.

The term "steel" actually refers to a wide variety of products. The various forms of steel all contain iron and carbon. They also contain one or more other elements, such as **silicon**, **titanium**, **vanadium**, **chromium**, **manganese**, cobalt, nickel, **zirconium**, **molybdenum**, and **tungsten**. Two other steel-like products are cast iron and wrought iron. Cast

Although now outdated, cast-iron stoves were once the primary source of heat for many homes, as well as a means for cooking. IMAGE COPYRIGHT 2009, TYLER OLSON. USED UNDER LICENSE FROM SHUTTERSTOCK.COM.

Long before digital cameras were invented, photographers printed pictures on metal plates. The tintype photograph (also known as a ferrotype), shown here featuring two African American soldiers during the Civil War (1861–1865), was actually made from a sheet of iron (not tin).

iron is an alloy of iron, carbon, and silicon. Wrought iron contains iron and any one or more of many other elements. In general, however, wrought iron tends to contain very little carbon.

Uses

It would be impossible to list all uses of iron and steel products. In general, those products can be classified into categories: (1) automotive; (2) construction; (3) containers, packaging, and shipping; (4) machinery and industrial equipment; (5) rail transportation; (6) oil and gas industries; (7) electrical equipment; and (8) appliances and utensils.

Compounds

Some iron is made into compounds. The amount is very small compared to the amount used in steel and other iron alloys. Probably the fastest growing use of iron outside of alloys is in water treatment systems. The terms ferric and ferrous refer to two different forms in which iron occurs in compounds. Some of the important iron compounds are:

- ferric acetate ($Fe(C_2H_3O_2)_3$): used in the dyeing of cloth
- ferric ammonium oxalate ($Fe(NH_4)_3(C_2O_4)_4$): blueprints
- ferric arsenate ($FeAsO_4$): insecticide
- ferric chloride ($FeCl_3$): water purification and sewage treatment systems; dyeing of cloth; coloring agent in paints; additive for animal feed; etching material for engraving, photography, and printed circuits
- ferric chromate ($Fe_2(CrO_4)_3$): yellow pigment (coloring) for paints and ceramics
- ferric hydroxide ($Fe(OH)_3$): brown pigment for coloring rubber; water purification systems
- ferric phosphate ($FePO_4$): fertilizer; additive for animal and human foods
- ferrous acetate ($Fe(C_2H_3O_2)_2$): dyeing of fabrics and leather; wood preservative

- ferrous gluconate ($Fe(C_6H_{11}O_7)_2$): dietary supplement in "iron pills"
- ferrous oxalate (FeC_2O_4): yellow pigment for paints, plastics, glass, and ceramics; photographic developer
- ferrous sulfate ($FeSO_4$): water purification and sewage treatment systems; catalyst in production of ammonia; fertilizer; herbicide; additive for animal feed; wood preservative; additive to flour to increase iron levels

Health Effects

Iron is of critical importance to plants, humans, and other animals. It occurs in hemoglobin, a molecule that carries oxygen in the blood. It picks up oxygen in the lungs and carries it to the cells. In the cells, oxygen is used to produce energy that the body needs to survive, grow, and stay healthy.

The U.S. Department of Agriculture (USDA) provides recommendations regarding the amount of vitamins and minerals that people should consume daily in order to stay healthy. Using a calculator on the USDA Web site, a person can determine how much iron he or she needs, based on the person's age, gender, height, weight, and physical activity level. Iron is found in a number of foods, including meat, eggs, and raisins.

An iron deficiency (lack of iron) can cause serious health problems in humans. For instance, hemoglobin molecules may not form in sufficient numbers. Or they may lose the ability to carry oxygen. If this occurs, a person develops a condition known as anemia. Anemia results in fatigue. Severe anemia can result in a lowered resistance to disease and an increase in heart and respiratory (breathing) problems. Some forms of anemia can even cause death.

Krypton

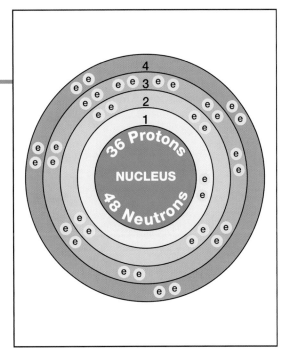

Overview

Krypton was one of three noble gases discovered in 1898 by Scottish chemist and physicist Sir William Ramsay (1852–1916) and English chemist Morris William Travers (1872–1961). Ramsay and Travers discovered the gases by allowing liquid air to evaporate. As it did so, each of the gases that make up normal air boiled off, one at a time. Three of those gases—krypton, **xenon**, and **neon**, were discovered for the first time this way.

The term noble gas refers to elements in Group 18 (VIIIA) of the periodic table. The periodic table is a chart that shows how chemical elements are related to each other. These gases have been given the name "noble" because they act as if they are "too arrogant" to react with other elements. Until the 1960s, no compound of these gases was known. Since they are so inactive, they are also called the inert gases. Inert means inactive.

Krypton has relatively few commercial uses. All of them involve lighting systems in one way or another.

Discovery and Naming

By 1898, two members of the noble gas family had been discovered. They were **helium** (atomic number 2) and **argon** (atomic number 18).

WORDS TO KNOW

Isotopes: Two or more forms of an element that differ from each other according to their mass number.

Noble gas: An element in Group 18 (VIIIA) of the periodic table.

Periodic table: A chart that shows how the chemical elements are related to each other.

Phosphor: Material that gives off light when struck by electrons.

Radioactive isotope: An isotope that breaks apart and gives off some form of radiation.

Spectrum (plural: spectra): The pattern of light given off by a glowing object, such as a star.

But no other elements in the family had been found. The periodic table contained empty boxes between helium and argon and below argon. The missing noble gases had atomic numbers 10, 36, 54, and 86. Chemists think of empty boxes in the periodic table as "elements waiting to be discovered."

Since the two known noble elements, helium and argon, are both gases, Ramsay and Travers hoped the missing elements were also gases. And if they were, they might be found in air. The problem was that air had already been carefully analyzed and found to be about 99.95 percent **oxygen**, **nitrogen**, and argon. Was it possible that the missing gases were in the last 0.05 percent of air?

To answer the question, the chemists worked not with air itself, but with liquid air. Air becomes liquid simply by cooling it far enough. The colder air becomes, the more gases within it turn into liquids. At −297.33°F (−182.96°C), oxygen changes from a gas into a liquid. At −320.42°F (−195.79°C), nitrogen changes from a gas into a liquid. And so on. Eventually, all the gases in air can be made to liquefy (change into a liquid).

But the reverse process also takes place. Suppose a container of liquid air holds 100 liters. The liquid air will warm up slowly. When its temperature reaches −320.42°F (−195.79°C), liquid nitrogen changes back to a gas. Since about 78 percent of air is nitrogen, only 22 percent of the original liquid air (22 liters) will be left.

When the temperature reaches −297.33°F (−182.96°C), oxygen changes from a liquid back to a gas. Since oxygen makes up 21 percent of air, another 21 percent (21 liters) of the liquid air will evaporate.

The work of Ramsay and Travers was very difficult, however, because the gases they were looking for are not abundant in air. Krypton, for example, makes up only about 0.000114 percent of air. For every 100 liters of liquid air, there would be only 0.00011, or about one-tenth of a milliliter of krypton. A 10th of a milliliter is about a drop. So Ramsay and Travers—although they didn't know it—were looking for one drop of krypton in 100 liters of liquid air!

Amazingly, they found it. The discovery of these three gases was a great credit to their skills as researchers. They suggested the name krypton for the new element. The name was taken from the Greek word *kryptos* for "hidden."

Physical Properties

Krypton is a colorless, odorless gas. It has a boiling point of –243.2°F (–152.9°C) and a density of 3.64 grams per liter. That makes krypton about 2.8 times as dense as air.

Chemical Properties

For many years, krypton was thought to be completely inert. Then, in the early 1960s, it was found to be possible to make certain compounds of the element. English chemist Neil Bartlett (1932–2008) found ways to combine noble gases with the most active element of all, **fluorine**. In 1963, the first krypton compounds were made—krypton difluoride (KrF_2) and krypton tetrafluoride (KrF_4). Other compounds of krypton have also been made since that time. However, these have no commercial uses. They are only laboratory curiosities.

Occurrence in Nature

The abundance of krypton in the atmosphere is thought to be about 0.000108 to 0.000114 percent. The element is also formed in Earth's crust when **uranium** and other radioactive elements break down. The amount in Earth's crust is too small to estimate, however.

Isotopes

Six naturally occurring isotopes of krypton exist. They are krypton-78, krypton-80, krypton-82, krypton-83, krypton-84, and krypton-86. Isotopes are two or more forms of an element. Isotopes differ from each other according to their mass number. The number written to the right of the

Krypton vs. Kryptonite

The famous cartoon character Superman has many super powers. Everybody knows that. He's the Man of Steel. He has X-ray vision. His hearing is so good, he can tune in on one voice in a crowded city. And, of course: "He's faster than a speeding bullet! More powerful than a loco-motive! Able to leap tall buildings in a single bound!"

But there's one substance that weakens Superman: kryptonite! If exposed to kryptonite, Superman experiences pain and loses his super powers. If exposed for too long, he can even die.

Kryptonite, of course, is purely fictional. Despite the similarity in names, kryptonite has nothing to do with element 36, krypton. According to

cartoon legend, Superman came from the planet Krypton. Kal-El, as he was originally known, was placed in a spaceship by his parents, moments before the planet exploded.

Unfortunately, as the young Superman blasted away from Krypton, a piece of kryptonite got stuck on the spaceship. The same terrible forces that caused the planet to explode, also had cre-ated the deadly kryptonite. And, as Superman would later find out, arch-villains always seem to get their hands on this green glowing rock!

Aside from the fictitious nature of kryptonite, there is another difference between it and krypton. Kryptonite is a rock—one that can cause great harm to Superman. Krypton is an inert gas that has no effect on anything.

element's name is the mass number. The mass number represents the number of protons plus neutrons in the nucleus of an atom of the element. The number of protons determines the element, but the number of neu-trons in the atom of any one element can vary. Each variation is an isotope.

Twenty-three radioactive isotopes of krypton are known also. A radio-active isotope is one that breaks apart and gives off some form of radiation. Radioactive isotopes are produced when very small particles are fired at atoms. These particles stick in the atoms and make them radioactive.

One radioactive isotope of krypton is used commercially, krypton-85. It can be combined with phosphors to produce materials that shine in the dark. A phosphor is a material that shines when struck by elec-trons. Radiation given off by krypton-85 strikes the phosphor. The phos-phor then gives off light. The same isotope is also used for detecting leaks in a container. The radioactive gas is placed inside the container to be tested. Since the gas is inert, krypton will not react with anything else in the container. But if the container has a leak, some radioactive krypton-85 will escape. The isotope can be detected with special devices for detecting radiation.

Cartoonists attempted to bring a little science into the Superman comics by having the superhero lose his powers when exposed to kryptonite. However, kryptonite is not related to krypton. In fact, it doesn't really exist. AP IMAGES.

Krypton-85 is also used to study the flow of blood in the human body. It is inhaled as a gas, and then absorbed by the blood. It travels through the bloodstream and the heart along with the blood. Its pathway can be followed by a technician who holds a detection device over the patient's body. The device shows where the radioactive material is going and how fast it is moving. A doctor can determine whether this behavior is normal or not.

Extraction

Krypton is still obtained by allowing liquid air to evaporate.

Uses

The only commercial uses of krypton are in various kinds of lamps. When an electric current is passed through krypton gas, it gives off a very

Krypton's Role in Metrics

The meter is the standard unit of length in the metric system. It was first defined in 1791. As part of the great changes brought by the French Revolution, an entirely new system of measurement was created: the metric system.

At first, the meter was defined in a very simple way. It was the distance between two lines scratched into a metal bar kept outside Paris. For many years, that definition was satisfactory for most purposes. Of course, it created a problem. Suppose someone in the United States was in the business of making meter sticks. That person would have to travel to Paris to make a copy of the official meter. Then the copy would have to be used to make other copies. The chances for error in this process are significant.

In 1960, scientists had another idea. They suggested using light produced by hot krypton as the standard of length. Here is how that standard was developed:

When an element is heated, it absorbs energy from the heat. The atoms present in the element are in an "excited," or energetic, state. Atoms normally do not remain in an excited state very long. They give off the energy they just absorbed and return to their normal, "unexcited" state.

The energy they give off can take different forms. One of those forms is light.

The kind of light given off is different for each element and for each isotope. The light usually consists of a series of very bright lines called a spectrum. The number and color of the lines produced is specific to each element and isotope.

When one isotope of krypton, krypton-86, is heated, it gives off a very clear, distinct, bright line with a reddish-orange color. Scientists decided to define the meter in terms of that line. They said that a meter is 1,650,763.73 times the width of that line.

This standard had many advantages. For one thing, people worldwide could find the official length of a meter. All one needed was the equipment to heat a sample of krypton-86. Then one had to look for the reddish-orange line produced. The length of the meter, then, was 1,650,763.73 times the width of that line.

This definition for the meter lasted only until 1983. Scientists then decided to define a meter by how fast light travels in a vacuum. This system is even more exact than the one based on krypton-86.

bright light. Perhaps the most common application of this principle is in airport runway lights. These lights are so bright that they can be seen even in foggy conditions for distances up to 1,000 feet (300 meters). The lights do not burn continuously. Instead, they send out very brief pulses of light. The pulses last no more than about 10 millionths of a second (10 microseconds). They flash on and off about 40 times per minute. Krypton is also used in slide and movie projectors.

Krypton gas is also used in making "neon" lights. Neon lights are colored lights often used in advertising. They are similar to fluorescent lightbulbs. But they give off a colored light because of the gas they contain. Some neon lights *do* contain the gas neon, but others contain other noble gases. A neon light filled with krypton, for example, glows yellow.

Compounds

Compounds of krypton have been prepared in the laboratory but do not exist in nature. The synthetic (artificial) compounds are used for research purposes only.

Health Effects

There is no evidence that krypton is harmful to humans, other animals, or plants.

Lanthanum

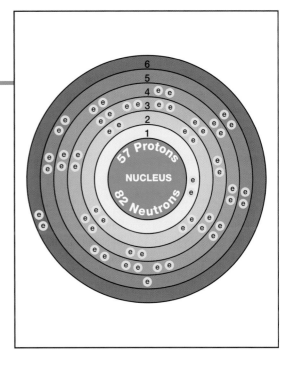

Overview

Lanthanum is the third element in Row 6 of the periodic table. The periodic table is a chart that shows how chemical elements are related to each other. Lanthanum is a transition metal in Group 3 (IIIB) of the periodic table. The transition metals are found in the center of the periodic table, in Groups 3 through 12.

Lanthanum can also be classified as a rare earth element. The rare earth elements include the 15 elements that make up Row 6 of the periodic table between barium and hafnium. That group of elements is also called the lanthanoids. Either way of classifying lanthanum is acceptable to most chemists.

Lanthanum was first discovered by Swedish chemist Carl Gustav Mosander (1797–1858) in 1839. This discovery was the first chapter in a long and interesting story. At the end of that story, six more elements had been discovered. All of these elements occur together in nature and were once hard to separate from each other.

Lanthanum metal has relatively few uses. Some of its compounds however, are used in lamps, color television sets, cigarette lighters, optical fibers, and hybrid cars.

Key Facts

Symbol: La

Atomic Number: 57

Atomic Mass: 138.90547

Family: Group 3 (IIIB); transition metal

Pronunciation: LAN-tha-num

WORDS TO KNOW

Alloy: A mixture of two or more metals with properties different from those of the individual metals.

Carbon arc lamp: A lamp for producing very bright white light.

Ductile: Capable of being drawn into thin wires.

Half life: The time it takes for half of a sample of a radioactive element to break down.

Isotopes: Two or more forms of an element that differ from each other according to their mass number.

Lanthanoids (rare earth elements): The elements that make up Row 6 of the periodic table between barium and hafnium.

Malleable: Capable of being hammered into thin sheets.

Periodic table: A chart that shows how the chemical elements are related to each other.

Phosphor: A material that gives off light when struck by electrons.

Radioactivity: Having a tendency to break down and give off radiation.

Discovery and Naming

Toward the end of the 1830s, Mosander became interested in an unusual black stone found near the town of Bastnas, Sweden. He learned that the stone contained two new materials. He thought those materials were both new elements. Mosander called them **cerium** and lanthanum. He was right about cerium, but wrong about lanthanum. The material Mosander called lanthanum later turned out to be a mixture of six new elements.

It took scientists more than 60 years to sort out these elements and separate them from each other. It was not until 1923 that a pure sample of lanthanum metal was even prepared. Still, Mosander is given credit for the discovery of lanthanum.

Physical Properties

Lanthanum is a white, ductile, malleable metal. Ductile means capable of being drawn into thin wires. Malleable means capable of being hammered into thin sheets. Lanthanum is relatively soft and can be cut with a sharp knife. Its melting point is 1,690°F (920°C) and its boiling point is 6,249°F (3,454°C). Its density is about 6.18 grams per cubic centimeter.

Magnified view of lanthanum aluminate crystal. © MICHAEL W. DAVIDSON, NATIONAL AUDUBON SOCIETY COLLECTION/PHOTO RESEARCHERS, INC.

Chemical Properties

Lanthanum is a very active metal. It reacts with most acids and with cold water, although slowly. With hot water, it reacts more quickly:

$$2La + 6H_2O \rightarrow 2La(OH)_3 + 3H_2$$

Lanthanum also reacts with **oxygen** in the air, especially if the air is moist.

Occurrence in Nature

Lanthanum is relatively common in Earth's crust. Its abundance is thought to be as high as 18 parts per million. That would make it nearly as common as **copper** or **zinc**. Unlike those metals, however, it usually does not occur in one place, as in copper mines. Instead, it is spread widely throughout Earth's crust. Its most common minerals are monazite, bastnasite, and cerite. These minerals generally contain all the other rare earth elements as well.

Isotopes

Two naturally occurring isotopes of lanthanum are known. They are lanthanum-138 and lanthanum-139. Isotopes are two or more forms of an element. Isotopes differ from each other according to their mass number. The number written to the right of the element's name is the mass number. The mass number represents the number of protons plus neutrons in the nucleus of an atom of the element. The number of protons determines the element, but the number of neutrons in the atom of any one element can vary. Each variation is an isotope.

Lanthanum-138 is very rare and is radioactive. Its half life is about 100 billion years. The half life of a radioactive element is the time it takes for half of a sample of the element to break down. Only 5 grams of a 10-gram sample of lanthanum-138 will remain after 100 billion years. The other 5 grams would have broken down to form a new isotope.

Thirty-six artificial radioactive isotopes have also been made. These isotopes are produced when very small particles are fired at atoms. These particles stick in the atoms and make them radioactive. None of the radioactive isotopes of lanthanum have any commercial use.

Extraction

The rare earth elements are very similar to each other. Separating them is a difficult task. The ores are first treated with sulfuric acid (H_2SO_4). The materials produced are then passed through a series of steps and the individual elements separated from each other.

Uses and Compounds

One of the most important uses of lanthanum compounds is in carbon arc lamps. In a carbon arc lamp, an electrical current is passed through the lamp electrode. The electrode is made of carbon and traces of other materials that have been added. The current causes the carbon to heat up and give off a brilliant white light. The exact color of the light depends on the other materials that have been added to the carbon. Lanthanum fluoride (LaF_3) and lanthanum oxide (La_2O_3) are usually used for this purpose.

These two compounds are also used in making phosphors. A phosphor is a material that gives off light when struck by electrons. The color of the phosphor depends on the elements present in the phosphor. The colors produced in a color television set are caused by phosphors painted on the back of the screen.

Compounds of lanthanum are also used to make special kinds of glass. High quality lenses, for example, are often made of glass containing a small amount of lanthanum.

One of the oldest uses of lanthanum metals is in the production of misch metal. Misch metal is an alloy that produces sparks when struck. One application of misch metal is in the manufacture of cigarette lighters.

A newer application of lanthanum glass is in making optical fibers. An optical fiber is a wire-like material made of glass. It carries light in the same way a copper wire carries electricity. Optical fibers have become popular methods for carrying audio, video, and digital messages. In many cases, optical fibers have replaced copper wires for this purpose.

In 1974, a team of French researchers invented a product called ZBLAN. It is made of **zirconium**, **barium**, lanthanum, **aluminum**, and **sodium**. ("ZBLAN" comes from the first letters of each element's symbol.) Scientists have found that ZBLAN is 100 times better at carrying messages than traditional optical fibers.

Lanthanum is also used in nickel-metal hydride (NiMH) batteries found in some hybrid cars.

As of 2010, China produced most of the rare earth elements used in high-tech and green (environmentally friendly) industries. Although such elements are found in other countries, including the United States, most nations rely on China to produce them. However, China's own growing demand for such elements raised concern about the potential for a rare earth supply shortage.

Health Effects

Lanthanum and its compounds are poisonous in high concentrations. They should be handled with care.

Lead

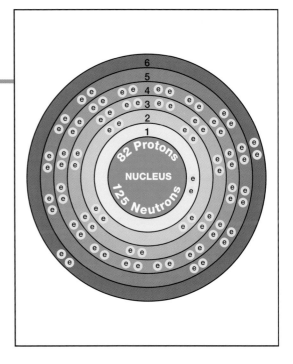

Overview

Lead is the heaviest member of the **carbon** family. The carbon family consists of the five elements in Group 14 (IVA) of the periodic table. The periodic table is a chart that shows how chemical elements are related to each other. Although a member of the carbon family, lead looks and behaves very differently from carbon.

Lead is one of only a few elements known to ancient peoples. One of the oldest examples of lead is a small statue found in Egypt. It was made during the First Dynasty, in about 3400 BCE. Mention of lead and lead objects can also be found in very old writings from India. And the Bible mentions lead in a number of passages.

Throughout history, lead has been used to make water and sewer pipes; roofing; cable coverings; type metal and other alloys; paints; wrappings for food, tobacco, and other products; and as an additive in gasoline. Since the 1960s, however, there has been a growing concern about the health effects of lead. For instance, scientists have found that lead can cause mental and physical problems in growing children. As a result, many common lead products have been or are being phased out.

Key Facts

Symbol: Pb

Atomic Number: 82

Atomic Mass: 207.2

Family: Group 14 (IVA); carbon

Pronunciation: LED

WORDS TO KNOW

Ductile: Capable of being drawn into thin wires.

Isotopes: Two or more forms of an element that differ from each other according to their mass number.

Laser: A device for making very intense light of one very specific color that is intensified many times over.

Malleable: Capable of being hammered into thin sheets.

Periodic table: A chart that shows how the chemical elements are related to each other.

Radioactive isotope: An isotope that breaks apart and gives off some form of radiation.

Toxic: Poisonous.

Discovery and Naming

Lead has been around for thousands of years. It is impossible to say when humans first discovered the element. It does not occur as an element in the earth very often. But one of its ores, lead sulfide (PbS), is fairly common. It is not difficult to obtain pure lead metal from lead sulfide. Humans probably discovered methods for doing so thousands of years ago.

By Roman times, lead metal was widely used. The far-reaching system that brought water to Rome contained many lead pipes. Sheets of lead were used as writing tablets and some Roman coins were also made of lead. Of particular interest was the use of lead in making pots and pans. Modern scientists believe many Romans may have become ill and died because of this practice. Cooking liquids in lead utensils tends to make the lead dissolve. It got into the food being cooked. People who ate those foods got more and more lead into their bodies. Eventually, the effects of lead poisoning must have begun to appear.

Of course, the Romans had little understanding of the connection between lead and disease. They probably never realized that they were poisoning themselves by using lead pots and pans.

No one is quite sure how lead got its name. The word has been traced to manuscripts that date to before the 12th century. Romans called the metal *plumbum*. It is from this name that the element's chemical symbol comes: Pb. Compounds of lead are sometimes called by this old name, such as plumbous chloride.

Physical Properties

Lead is a heavy, soft, gray solid. It is both ductile and malleable. Ductile means capable of being drawn into thin wires. Malleable means capable of being hammered into thin sheets. It has a shiny surface when first cut, but it slowly tarnishes (rusts) and becomes dull. Lead is easily worked. "Working" a metal means bending, cutting, shaping, pulling, and otherwise changing the shape of the metal.

The melting point of lead is 621.3°F (327.4°C), and its boiling point is 3,180 to 3,190°F (1,750 to 1,755°C). Its density is 11.34 grams per cubic centimeter. Lead does not conduct an electric current, sound, or vibrations very well.

Chemical Properties

Lead is a moderately active metal. It dissolves slowly in water and in most cold acids. It reacts more rapidly with hot acids. It does not react with **oxygen** in the air readily and does not burn.

Occurrence in Nature

The abundance of lead in Earth's crust is estimated to be between 13 and 20 parts per million. It ranks in the upper third among the elements in terms of its abundance.

Lead rarely occurs as a pure element in the earth. Its most common ore is galena, or lead sulfide (PbS). Other ores of lead are anglesite, or lead sulfate ($PbSO_4$); cerussite, or lead carbonate ($PbCO_3$); and mimetite ($PbCl_2 \cdot Pb_3(AsO_4)_2$).

As of 2008, the largest producers of lead resources in the world included China, Australia, the United States, Peru, Mexico, Canada, and India. In the United States, most of the lead produced came from Missouri, Alaska, Idaho, Montana, and Washington. According to the U.S. Geological Survey (USGS), 88 percent of the lead used in the United States was used by the lead-acid storage battery industry.

Isotopes

Four naturally occurring isotopes of lead occur. They are lead-204, lead-206, lead-207, and lead-208. Isotopes are two or more forms of an element. Isotopes differ from each other according to their mass number. The number written to the right of the element's name is the mass number. The mass number represents the number of protons plus neutrons in the nucleus of an atom of the element. The number of protons determines the element, but the number of neutrons in the atom of any one element can vary. Each variation is an isotope.

One naturally occurring isotope of lead is radioactive, lead-204. In addition, 44 radioactive isotopes of lead have been made artificially in the laboratory. A radioactive isotope is one that breaks apart and gives off some form of radiation. Radioactive isotopes are produced when very small particles are fired at atoms. These particles stick in the atoms and make them radioactive.

One radioactive isotope of lead, lead-210, is sometimes used in medicine. This isotope gives off radiation that can kill cancer cells. It is also used to treat non-cancerous eye disorders.

Extraction

Lead is obtained from its ores by a method used with many metals. First, the ore is roasted (heated in air). Roasting, also called smelting, converts the ore to a compound of lead and oxygen, lead oxide (PbO_2). Lead oxide is then heated with charcoal (pure carbon). The carbon takes oxygen away from the lead oxide. It leaves pure lead behind:

$$PbO_2 + C \xrightarrow{\text{heated}} CO_2 + Pb$$

Lead obtained in this way is not very pure. It can be purified electrolytically. Electrolytic refining involves passing an electric current through a compound. Very pure lead is collected at one side of the container in which the reaction is carried out.

Lead is also recovered in recycling programs. Recycling is the process by which a material is retrieved from a product that is no longer used. For example, old car batteries were once just thrown away. Now they are sent to recycling plants where lead can be extracted and used over and over again. It is not necessary to get all the lead that industry needs from new sources, such as ores.

Uses

The lead industry is undergoing dramatic change. Many products once made with lead no longer use the element. The purpose of this change is to reduce the amount of lead that gets into the environment. Examples of such products include ammunition, such as shot and bullets; sheet lead used in building construction; solder; water and sewer pipes; ball bearings; radiation shielding; and gasoline. These changes are possible because manufacturers are finding safer elements to use in place of lead.

Other uses of lead have not declined. The best example is lead storage batteries. A lead storage battery is a device for converting chemical energy into electrical energy. The majority of cars and trucks has at least one lead storage battery. But no satisfactory substitute for it has been found. About 88 percent of all lead consumed in the United States goes to the manufacture of lead storage batteries. In addition to cars and trucks, these batteries are used for communication networks and emergency power supplies in hospitals, and in forklifts, airline ground equipment, and mining vehicles.

Compounds

A small percentage of lead is used to make lead compounds. Although the amount of lead is small, the variety of uses for these compounds is large. Some examples of important lead compounds are:

- lead acetate $(Pb(C_2H_3O_2)_2)$: insecticides; waterproofing; varnishes; dyeing of cloth; production of gold; hair dye
- lead antimonate $(Pb_3(SbO_4)_2)$: staining of glass, porcelain and other ceramics
- lead azide $(Pb(N_3)_2)$: used as a "primer" for high explosives
- lead chromate ("chrome yellow"; $PbCrO_4$): industrial paints (use restricted by law)
- lead fluoride (PbF_2): used to make lasers; specialized optical glasses
- lead iodide (PbI_2): photography; cloud seeding to produce rain
- lead naphthenate $(Pb(C_7H_{12}O_2))$: wood preservative; insecticide; additive for lubricating oil; paint and varnish drier
- lead phosphite $(2PbO \cdot PbHPO_3)$: used to screen out ultraviolet radiation in plastics and paints
- lead stearate $(Pb(C_{18}H_{35}O_2)_2)$: used to make soaps, greases, waxes, and paints; lubricant; drier for paints and varnishes
- lead telluride $(PbTe)$: used to make semiconductors, photoconductors, and other electronic equipment

Health Effects

The health effects of lead have become much better understood since the middle of the 20th century. At one time, the metal was regarded as quite

The High Cost of Lead Gasoline

For many years, lead was regarded as a miracle chemical by the automotive industry. The power to run gas-fueled cars comes from the burning of gasoline in the engine. However, burning gasoline is not a simple process. Many things happen inside an engine when gasoline burns in the carburetor.

For example, an engine can "knock" if the gasoline does not burn properly. "Knocking" is a "bang-bang" sound from the engine. It occurs when low-grade gasoline is used.

One way to prevent knocking is to use high-grade gasoline. Another way is to add chemicals to the gasoline. The best gasoline additive discovered was a compound called tetraethyl lead $(Pb(C_2H_5)_4$. Tetraethyl lead was usually called "lead" by the automotive industry, the consumer, and advertisers. When someone bought "leaded" gasoline, it contained not lead metal, but tetraethyl lead.

Leaded gasoline was a great discovery. It could be made fairly cheaply and it prevented car engines from knocking. No wonder people thought it was a miracle chemical.

What people didn't realize was that tetraethyl lead breaks down in a car engine because of the high temperature at which engines operate. When tetraethyl lead breaks down, elemental lead (Pb) is formed:

$$Pb(C_2H_5)_4 \xrightarrow{\text{heat}} Pb + \text{other products}$$

The result—with millions of cars being driven every day—was more and more lead getting into the air. And more and more people inhaled that lead. Eventually, doctors began to see more people with lead-related diseases.

The U.S. government finally decided that tetraethyl lead was too dangerous to use in gasoline. By 1990, the use of this compound had been banned by all governments in North America.

safe to use for most applications. Now lead is known to cause both immediate and long-term health problems, especially with children. It is toxic when swallowed, eaten, or inhaled.

Young children are most at risk from lead poisoning. Some children have a condition known as pica. They have an abnormal desire to eat materials like dirt, paper, and chalk. Children with pica sometimes eat paint chips off walls. At one time, many interior house paints were made with lead compounds. Thus, crawling babies or children with pica ran the risk of eating large amounts of lead and being poisoned.

Some symptoms of lead poisoning include nausea, vomiting, extreme tiredness, high blood pressure, and convulsions (spasms). Over a long period of time, these children often suffer brain damage. They lose the ability to carry out normal mental functions.

According to the U.S. Environmental Protection Agency (EPA), the most common source of lead in homes built prior to 1978 is lead-based paint. When paint peels, some of it becomes airborne and could be ingested. Also, some young children chew on chips of paint, not realizing the dangers of lead poisoning. IMAGE COPYRIGHT 2009, JOY FERA. USED UNDER LICENSE FROM SHUTTERSTOCK.COM.

Other forms of lead poisoning can also occur. For example, people who work in factories where lead is used can inhale lead fumes. The amount of fumes inhaled at any one time may be small. But over months or years, the lead in a person's body can build up. This type of lead poisoning can lead to nerve damage and problems with the gastrointestinal system (stomach and intestines).

Today, there is an effort to reduce the use of lead in consumer products. For instance, older homes are often tested for lead paint before they are resold. Lead paint has also been removed from older school buildings.

However, in recent years, various toys and other products imported from other countries have been found to contain lead paint. This situation has prompted a number of recalls. The U.S. Centers for Disease Control and Prevention (CDC; www.cdc.gov) maintains of list of products being recalled due to lead concerns.

Lithium

Overview

Lithium is the first member of the alkali metal family. The alkali metals are the elements that make up Group 1 (IA) of the periodic table. The periodic table is a chart that shows how chemical elements are related to one another. The alkali metals include **sodium**, **potassium**, **rubidium**, **cesium**, and **francium**. Lithium is also the least dense of all metals. It has a density about half that of water.

Credit for the discovery of lithium usually goes to Swedish chemist Johan August Arfwedson (or Arfvedson; 1792–1841). Arfwedson found the new element in a mineral that had first been identified about 20 years earlier by Brazilian scientist Jozé Bonifácio de Andrada e Silva (1763–1838). That mineral, petalite, is still a major source of lithium today.

Lithium has a number of important and interesting uses. In recent years, it has been used to make lightweight, efficient batteries. Compounds of lithium have also been used to treat a mental disorder known as bipolar disorder.

Discovery and Naming

The first clues to the existence of lithium surfaced in 1800. De Andrada was a Brazilian scientist and statesman visiting in Scandinavia. During

Key Facts

Symbol: Li

Atomic Number: 3

Atomic Mass: [6.941]

Family: Group 1 (IA); alkali metal

Pronunciation: LI-thee-um

WORDS TO KNOW

Alkali metals: Elements that make up Group 1 (IA) of the periodic table.

Alloy: A mixture of two or more metals with properties different from those of the individual metals.

Bipolar disorder: A condition in which a person experiences wild mood swings.

Catalyst: A substance used to speed up a chemical reaction without undergoing any change itself.

Isotopes: Two or more forms of an element that differ from each other according to their mass number.

Periodic table: A chart that shows how the chemical elements are related to each other.

Radioactive isotope: An isotope that breaks apart and gives off some form of radiation.

one of his trips to the countryside, he came across a mineral that he did not recognize. He called the mineral petalite.

Some scientists were not convinced that petalite was a new mineral. But in 1817, the same mineral was rediscovered on the island of Utö. Interest in the mineral grew.

Arfwedson was troubled by the results of his analysis of petalite. In his studies, he could not identify 10 percent of the mineral. He finally concluded that the missing 10 percent must be a new element. He called the new element lithium, from the Greek word *lithos* for "stone."

Arfwedson was not able to produce pure lithium. About a year later, however, Swedish chemist William Thomas Brande (1788–1866) and English chemist Sir Humphry Davy (1778–1829) were both able to extract the pure metal from its compounds.

Physical Properties

Lithium is a very soft, silvery metal. It has a melting point of 356.97°F (180.54°C) and a boiling point of about 2,435°F (1,335°C). Its density is 0.534 grams per cubic centimeter. By comparison, the density of water is 1.000 grams per cubic centimeter. Lithium's hardness on the Mohs scale is 0.6. The Mohs scale is a way of expressing the hardness of a material. It runs from 0 (for talc) to 10 (for diamond). A hardness of 0.6 means that the material can be scratched with a fingernail.

Chemical Properties

Lithium is an active element, but not as active as the other alkali metals. It reacts slowly with water at room temperature and more rapidly at higher temperatures. It also reacts with most acids, giving off **hydrogen** gas. Lithium does not react with **oxygen** at room temperature, but above 212°F (100°C) does so to form lithium oxide (Li_2O). Under the proper conditions, the element also combines with **sulfur**, hydrogen, **nitrogen**, and the halogens.

Occurrence in Nature

The abundance of lithium in Earth's crust is estimated to be about 0.005 percent. That places it among the top 15 elements found in the earth. The most common ores of lithium are spodumene, petalite, and lepidolite. Lithium is also obtained from saltwater. As saltwater evaporates, dissolved solids are left behind. These solids include sodium chloride (NaCl), potassium chloride (KCl), and lithium chloride (LiCl).

In 2008, the world's largest consumer of lithium minerals and compounds was the United States. The major producer of lithium chemicals worldwide was Chile. Other countries involved in lithium production included Argentina, Australia, Brazil, Canada, China, Portugal, the United States, and Zimbabwe. Specific information on U.S. production was not released in order to preserve trade secrets.

Isotopes

Two naturally occurring isotopes of lithium exist: lithium-6 and lithium-7. Isotopes are two or more forms of an element. Isotopes differ from each other according to their mass number. The number written to the right of the element's name is the mass number. The mass number represents the number of protons plus neutrons in the nucleus of an atom of the element. The number of protons determines the element, but the number of neutrons in the atom of any one element can vary. Each variation is an isotope.

In addition, five radioactive isotopes of lithium have been produced. A radioactive isotope is one that breaks apart and gives off some form of radiation. Radioactive isotopes are produced when very small particles are fired at atoms. These particles stick in the atoms and make them radioactive. None of these isotopes has any important commercial application.

Extraction

Lithium compounds are first converted to lithium chloride (LiCl). Then, an electric current is passed through molten (melted) lithium chloride. The current separates the compound into lithium and **chlorine** gas:

$$2LiCl \longrightarrow \text{electric current} \rightarrow 2Li + Cl_2$$

Uses and Compounds

One of the primary uses of lithium metal is in the production of alloys. An alloy is made by melting and mixing two or more metals. The mixture has properties different from those of the individual metals. The most common alloys of lithium are those with **magnesium** and **aluminum**. These alloys combine the properties of strength and lightness. One lithium alloy, called Bahnmetall, is used to make very strong ball bearings for machinery, such as the wheel bearings in railroad cars. An alloy of lithium and magnesium is used in space vehicles because it is strong, but very lightweight.

Lithium metal is also a component of certain kinds of batteries. A battery is a device for converting chemical energy into electrical energy. Lithium batteries are much lighter than the familiar **lead** and sulfuric acid batteries used in many cars and trucks. They also reduce the use of toxic lead and **cadmium**. Lithium batteries are used in products such as

Lithium-ion battery and charger. IMAGE COPYRIGHT 2009, MR. BRIGHTSIDE. USED UNDER LICENSE FROM SHUTTERSTOCK.COM.

Lithium and Mood Swings

A new use for lithium carbonate was discovered in 1949 when John Cade (1912–1980), an Australian physician, found that patients with bipolar disorder benefited from taking the substance. Bipolar disorder is a condition once known as manic-depressive disorder. The condition is characterized by dramatic mood swings. A person can be very happy and carefree one moment, but terribly depressed the next moment. Some patients become so depressed that they commit suicide. Until 1949, there was no effective treatment for bipolar disorder.

Cade found that most patients who took lithium carbonate were relieved of at least some of their symptoms. Their "high" points were not as high, and their "low" points were not as low. The compound helped someone with bipolar disorder to live a quieter, more normal life. Today, more than 60 percent of those with bipolar disorder benefit from lithium treatments. It is estimated that some 5.7 million American adults have bipolar disorder; however, children as young as six have been diagnosed with the disorder.

As with most medications, lithium compounds can have side effects. They can cause nausea, dizziness, diarrhea, dry mouth, and weight gain. But these side effects can usually be controlled. And they are often a small price to pay for relief from the severe effects of bipolar disorder.

watches, microcomputers, cameras, small appliances, electronic games, toys, and many types of military and space vehicles. Although the batteries tend to be more expensive than other kinds of batteries, they have the advantages of producing a constant supply of electrical energy and having a long shelf-life.

Compounds of lithium have a number of important uses. Two of the most significant applications are in the glass and ceramics field and in the production of aluminum. The addition of a small amount of lithium carbonate (Li_2CO_3) to a glass or ceramic makes the material stronger. Examples of the use of lithium carbonate are shock-resistant cookware and black-and-white television tubes.

Producers of aluminum also use lithium carbonate in preparing aluminum metal from aluminum oxide. Lithium carbonate reduces the heat needed to make the reaction occur. As a result, producers save money by using less energy.

Another important compound of lithium is lithium stearate. Lithium stearate is added to petroleum to make a thick lubricating grease. The grease is used in many industrial applications because it does not break down at high temperatures, it does not become hard when cooled, and it does not react with water or oxygen in the air. Lithium greases are used in military, industrial, automotive, aircraft, and marine applications. Lithium stearate is also used as an additive in cosmetics and plastics.

Lithium compounds are also used as catalysts in many different industrial processes. A catalyst is a substance used to speed up a chemical reaction. The catalyst does not undergo any change itself during the reaction. For example, one lithium catalyst is used to make tough, strong, synthetic (artificial) rubber. It does not have to be vulcanized (heat-treated) like natural rubber.

Health Effects

Lithium and its compounds have a range of effects on the human body. For instance, compounds of lithium tend to harm the kidneys. And lithium carbonate (Li_2CO_3) can affect a person's mental health, as shown in a discovery made in 1949 (see sidebar on page 319).

Lutetium

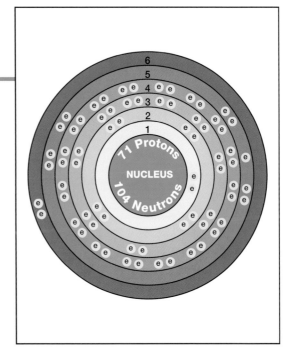

Inside the image: 6 5 4 3 2 1 with e (electrons) on each shell, 71 Protons, NUCLEUS, 104 Neutrons

Overview

Lutetium is the heaviest, rarest, and most expensive lanthanoid element. The lanthanoids elements make up Row 6 of the periodic table between barium and hafnium. The periodic table is a chart that shows how chemical elements are related to one another. The lanthanoids are usually shown as a separate row at the bottom of the table. They are also called the rare earth elements. That name does not fit very well for most lanthanoids. They are not really so rare, but were once difficult to separate from each other. However, lutetium is both rare and difficult to separate from the other lanthanoids.

Lutetium was first discovered in the early 1900s by two chemists working independently. It was found in a complex black mineral that had been found near the town of Ytterby, Sweden, in 1787.

Today, there are very few uses for lutetium metal.

Discovery and Naming

In 1787, a Swedish army officer, Carl Axel Arrhenius (1757–1824), found an odd black rock outside the town of Ytterby, Sweden. He gave the rock to a chemist friend, Johan Gadolin (1760–1852), for study.

Key Facts

Symbol: Lu

Atomic Number: 71

Atomic Mass: 174.9668

Family: Lanthanoid (rare earth metal)

Pronunciation: loo-TEE-she-um

WORDS TO KNOW

Catalyst: A substance used to speed up a chemical reaction without undergoing any change itself.

Ductile: Capable of being drawn into a thin wire.

Isotopes: Two or more forms of an element that differ from each other according to their mass number.

Lanthanoids: The elements in the periodic table with atomic numbers 57 through 71.

Radioactive isotope: An isotope that breaks apart and gives off some form of radiation.

Toxic: Poisonous.

That rock turned out to contain one of the most complex and interesting minerals ever discovered: yttria. Chemists kept busy for the next century trying to figure out exactly what materials were in yttria.

Eventually, they found nine new elements that had never been seen before. Separating these elements from each other was very difficult, however. The nine elements are chemically similar and all behave in nearly the same way. It is very difficult to know whether a sample of yttria contains one, two, three … or all nine of the elements.

In 1879 French chemist Jean-Charles-Galissard de Marignac (1817–1894) announced the discovery of a new element in yttria. He called the element **ytterbium**. Other chemists suspected that ytterbium was really a mixture of elements. They searched for ways to separate ytterbium into simpler parts.

It took nearly 30 years to solve this puzzle. And the answer came from three laboratories at nearly the same time. The first to report his results was French chemist Georges Urbain (1872–1938). In 1907, Urbain reported that ytterbium was not an element, but a mixture of two new elements. He called those elements neoytterbium and lutecium. The first name meant "new ytterbium." The second name comes from Lutecia, the ancient name for the city of Paris in France.

At nearly the same time, German chemist Karl Auer (Baron von Welsbach; 1858–1929) made the same discovery. He suggested different names for the two new elements in ytterbium. He called them cassiopeium and aldebaranium, in honor of the constellation Cassiopeia and the bright star Aldebaran. Today, some German chemists still refer to lutetium as cassiopeium.

A third chemist working on ytterbium was American chemist Charles James (1880–1926). James announced his discoveries after Urbain and Auer. Some authorities give credit for the discovery of lutetium to all three scientists.

None of these early scientists actually saw pure lutetium. The substance they thought was an element was actually a compound, usually lutetium oxide. The pure metal was isolated only quite recently.

In 1949, the spelling of the element changed from "lutecium" to "lutetium."

Physical Properties

Lutetium is a silvery white metal that is quite soft and ductile. The term ductile means capable of being drawn into thin wires. It has a melting point of 3,006°F (1,652°C) and a boiling point of 6,021°F (3,327°C). Its density is 8.49 grams per cubic centimeter.

Chemical Properties

Lutetium reacts slowly with water and dissolves in acids. Other chemical properties tend to be of interest only to researchers.

Occurrence in Nature

Lutetium is thought to be very rare in Earth's crust. It occurs to the extent of about 0.8 to 1.7 parts per million. That still makes it somewhat more common than better known elements such as **iodine**, **silver**, and **mercury**. The most common ore of lutetium is monazite, in which its concentration is about 0.003 percent.

Isotopes

There are two naturally occurring isotopes of lutetium: lutetium-175 and lutetium-176. Isotopes are two or more forms of an element. Isotopes differ from each other according to their mass number. The number written to the right of the element's name is the mass number. The mass number represents the number of protons plus neutrons in the nucleus of an atom of the element. The number of protons determines the element, but the number of neutrons in the atom of any one element can vary. Each variation is an isotope.

The second of these isotopes, lutetium-176, is radioactive. A radioactive isotope is one that breaks apart and gives off some form of radiation. Some radioactive isotopes occur in nature. Others can be produced by firing very small particles at atoms. These particles stick in the atoms and make them radioactive.

In addition, 55 radioactive isotopes have also been produced artificially. None of these isotopes has any commercial use.

Extraction

Lutetium is the most difficult lanthanoid to obtain in pure form. The usual method used begins with either lutetium fluoride (LuF_3) or lutetium chloride ($LuCl_3$). An active metal, such as **sodium** (Na) or **potassium** (K) is then added to LuF_3 or $LuCl_3$ to obtain pure lutetium. For example:

$$LuF_3 + 3Na \xrightarrow{\text{heated}} 3NaF + Lu$$

Uses

Lutetium is one of the most expensive lanthanoids. In 2007, lutetium oxide sold for about \$3,500 per kilogram. It is sometimes used as a catalyst in the petroleum industry. A catalyst is a substance used to speed up a chemical reaction. The catalyst does not undergo any change itself during the reaction. There are virtually no other uses for lutetium.

Compounds

There are no commercially important lutetium compounds.

Health Effects

The health effects of lutetium are not well known. Exposure to compounds of lutetium is thought to cause irritation of the skin, eyes, and respiratory (breathing) system.

Magnesium

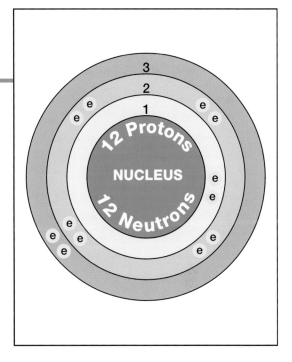

Overview

Magnesium is the second element in Group 2 (IIA) of the periodic table, a chart that shows how chemical elements are related to each other. The elements in Group 2 are known as the alkaline earth elements. Other elements in that group include **beryllium**, **calcium**, **strontium**, **barium**, and **radium**.

Compounds of magnesium have been used by humans for centuries. Yet, the element itself was not isolated until 1808. The long delay occurred because magnesium forms very stable compounds. That means that such compounds do not break down very easily.

Magnesium is the seventh most abundant element in Earth's crust. It also occurs in large amounts dissolved in ocean waters.

Large amounts of magnesium are used to make alloys. An alloy is made by melting or mixing two or more metals. The mixture has properties different from those of the individual metals. Magnesium alloys are quite light, yet very strong. This property makes them useful in the construction of airplanes and spacecraft.

About 60 percent of the magnesium compounds produced in the United States are used in the manufacture of refractory materials.

Key Facts

Symbol: Mg

Atomic Number: 12

Atomic Mass: 24.3050

Family: Group 2 (IIA); alkaline earth metal

Pronunciation: mag-NEE-zee-um

WORDS TO KNOW

Alkaline earth metal: An element found in Group 2 (IIA) of the periodic table.

Alloy: A mixture of two or more metals with properties different from those of the individual metals.

Enzyme: Catalyst in a living organism that speeds up the rate at which certain changes take place in the body.

Fabrication: Shaping, molding, bending, cutting, and working with a metal.

Isotopes: Two or more forms of an element that differ from each other according to their mass number.

Periodic table: A chart that shows how the chemical elements are related to each other.

Radioactive isotope: An isotope that breaks apart and gives off some form of radiation.

Refractory: A material that can withstand very high temperatures and reflects heat from itself.

A refractory material is one that can withstand very high temperatures by reflecting heat. Refractory materials are used to line the ovens that maintain high temperatures. Agriculture, construction, industrial, and chemical operations use portions of the remaining 40 percent.

Discovery and Naming

Compounds of magnesium are very abundant in the earth. Dolomite, or calcium magnesium carbonate ($CaMg(CO_3)_2$), is an example. Dolomite has been used as a building material for centuries.

Another well-known magnesium compound is Epsom salts, or magnesium sulfate ($MgSO_4$). Epsom salts are known for their soothing qualities, most notably when added to a bath.

Careful studies of magnesium and its compounds began in the middle 1700s. Scottish physician and chemist Joseph Black (1728–1799) carried out some of the earliest experiments on magnesium compounds. He reported on his research in an article that became famous. Black is sometimes given credit for "discovering" magnesium because of his work with the element.

By 1800, chemists knew that magnesium was an element. But no one had been able to prepare pure magnesium metal. Magnesium holds very tightly to other elements in its compounds. No one had found a way to break the bonds between magnesium and these other elements.

In 1808, English chemist Humphry Davy (1778–1829) solved the problem by passing an electric current through molten (melted) magnesium oxide (MgO). The current caused the compound to break apart, forming magnesium metal and **oxygen** gas:

$$2MgO \xrightarrow{\text{electric current}} 2Mg + O_2$$

Davy used this method to discover a number of other elements. Like magnesium, these elements form compounds that are very difficult to break apart. An electric current provides the energy to break these compounds down into their elements.

The name magnesium goes back many centuries. It was selected in honor of a region in Greece known as Magnesia. The region contains large supplies of magnesium compounds.

Physical Properties

Magnesium is a moderately hard, silvery-white metal. It is the lightest of all structural metals. These metals are strong enough to be used in constructing buildings, bridges, automobiles, and airplanes.

Magnesium is easily fabricated. Fabrication means shaping, molding, bending, cutting, and working with a metal. Metals must be fabricated before they can be turned into useful products. Metals that are strong, tough, or hard are not easily fabricated. They must be converted to an alloy. A metal that *is* more easily fabricated (such as magnesium) is combined with them.

The melting point of magnesium is 1,200°F (651°C) and its boiling point is 2,000°F (1,100°C). Its density is 1.738 grams per cubic centimeter.

Chemical Properties

Magnesium is a fairly active metal. It reacts slowly with cold water and more rapidly with hot water. It combines with oxygen at room temperature to form a thin skin of magnesium oxide. It burns with a blinding white light at higher temperatures. Magnesium reacts with most acids and with some alkalis. An alkali is a chemical with properties opposite those of an acid. Sodium hydroxide (common lye) and limewater are examples of alkalis.

Magnesium also combines easily with many non-metals, including **nitrogen**, **sulfur**, **phosphorus**, **chlorine**, **fluorine**, **bromine**, and **iodine**.

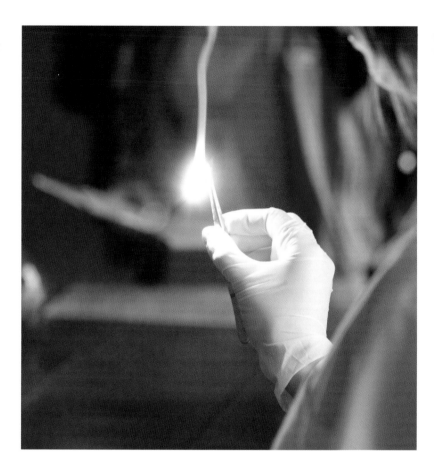

It also reacts readily with a number of compounds, such as carbon monoxide (CO), carbon dioxide (CO_2), sulfur dioxide (SO_2), and nitric oxide (NO).

Occurrence in Nature

The abundance of magnesium in Earth's crust is estimated to be about 2.1 percent. That makes it the sixth most common element in the earth. It also occurs in seawater. A cubic mile of seawater is estimated to contain up to six million tons of magnesium.

There are many naturally occurring minerals of magnesium. Some of the most important are dolomite; magnesite, or magnesium carbonate ($MgCO_3$); carnallite, or potassium magnesium chloride ($KMgCl_3$); and epsomite, or magnesium sulfate ($MgSO_4$).

In 2008, the largest producer of magnesium metal was reported to be China. Other large producers included Brazil, Israel, Kazakhstan, and Russia. The largest producers of magnesite were China, Turkey, North Korea, Russia, and Austria. The amounts of magnesium metal and magnesite produced in the United States were not reported in order to protect trade secrets.

Magnesium metal produced in the United States comes from three sources: seawater, brine, and mines. According to the U.S. Geological Survey (USGS), magnesium was recovered from brines in Utah's Great Salt Lake.

Isotopes

There are three naturally occurring isotopes of magnesium: magnesium-24, magnesium-25, and magnesium-26. Isotopes are two or more forms of an element. Isotopes differ from each other according to their mass number. The number written to the right of the element's name is the mass number. The mass number represents the number of protons plus neutrons in the nucleus of an atom of the element. The number of protons determines the element, but the number of neutrons in the atom of any one element can vary. Each variation is an isotope.

Twelve radioactive isotopes of magnesium also exist. A radioactive isotope is one that breaks apart and gives off some form of radiation. Radioactive isotopes are produced when very small particles are fired at atoms. These particles stick in the atoms and make them radioactive.

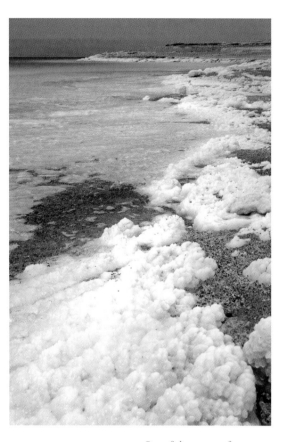

One of the sources of magnesium is seawater. It is processed in various locations in the United States and the world. Pictured here is the Dead Sea, which is bordered by Israel and Jordan in the Middle East. Much saltier than the ocean, the Dead Sea is rich in magnesium. IMAGE COPYRIGHT 2009, MYTHO. USED UNDER LICENSE FROM SHUTTERSTOCK.COM.

Extraction

Magnesium is prepared by one of two methods. The first method is similar to the method used by Davy in 1808. An electric current is passed through molten (melted) magnesium chloride:

$$2MgCl \xrightarrow{\text{electric current}} 2Mg + Cl_2$$

Magnesium alloys are used in the production of skis. IMAGE COPYRIGHT 2009, MONKEY BUSINESS IMAGES. USED UNDER LICENSE FROM SHUTTERSTOCK.COM.

The second method involves reacting magnesium oxide with ferrosilicon. Ferrosilicon is an alloy of **iron** and **silicon**. When magnesium oxide and ferrosilicon react, free magnesium metal is formed.

Uses

Although most cameras now use electronic flashes, magnesium metal is often contained in cameras that use flash bulbs. A thin strip of magnesium metal is inside the bulb. When the flash is ignited, the magnesium strip catches fire. It burns with a very bright white light. The light from the bulb illuminates a scene for the photograph.

A common use of magnesium metal is in fireworks. Most firework displays include some brilliant flashes of very white light. Those flashes are produced by the burning of magnesium metal.

Magnesium is commonly alloyed with other metals. Magnesium and **aluminum**, for instance, are two metals that combine to form alloys that are very strong and resistant to corrosion (rust). But they weigh much less than steel alloys with similar properties.

Strength and low density are important properties in the manufacture of airplanes, automobiles, metal luggage, ladders, shovels and other gardening equipment, racing bikes, skis, race cars, cameras, and power tools. A typical magnesium alloy contains about 90 percent magnesium, 2 to 9 percent aluminum, and small amounts of **zinc** and **manganese**.

Compounds

The largest single use of magnesium compounds is in refractories.

Other magnesium compounds are used in the following categories:

- Medicine: pain killer and fever reducer (magnesium acetylsalicylate); antacid to neutralize stomach acid (magnesium hydroxide; magnesium phosphate; magnesium silicate); laxative to loosen the bowels

The Popularity of Epsom Salts

Perhaps the best known magnesium compound is magnesium sulfate ($MgSO_4$). It is popularly known as Epsom salts.

One of the earliest stories about Epsom salts dates back to 1618. The town of Epsom, in Surrey, England, was suffering from a severe drought. A farmer named Henry Wicker brought his cattle to drink from a watering hole on the town commons (central park). But the cattle would not drink the water. Wicker was surprised because he knew they were very thirsty. He tasted the water himself and found that it was very bitter.

The bitterness was due to magnesium sulfate in the water. This compound became known as Epsom salts.

People soon learned that soaking in the natural waters that contained Epsom salts made them feel better. The salts seemed to have properties that soothed the body. Before long, soaking in these waters became very popular.

Today, Epsom salts can be used in bath water. They relax sore muscles and remove rough skin. Many people believe the salts have the same relaxing effect as hot springs.

Some gardeners even believe that sprinkling Epsom salts in the garden helps flowers and vegetables grow!

(magnesium carbonate; magnesium chloride; magnesium citrate; magnesium hydroxide; magnesium lactate; magnesium phosphate); antiseptic to kill germs (magnesium borate; magnesium salicylate; magnesium sulfate); sedative to help one sleep (magnesium bromide)

- Production of glass and ceramics: magnesium fluoride; magnesium oxide
- Mothproofing of textiles: magnesium hexafluorosilicate
- Fireproofing wood for construction: magnesium phosphate
- Manufacture of paper: magnesium sulfite

Health Effects

Magnesium is essential for good health in both plants and animals. It forms part of the chlorophyll molecule found in all green plants. Chlorophyll is the molecule in green plants that controls the conversion of carbon dioxide and water to carbohydrates, such as starch and sugars. Plants that do not get enough magnesium cannot make enough chlorophyll. Their leaves develop yellowish blotches as a result.

Magnesium is found in many enzymes in both plants and animals. An enzyme is a catalyst in a living organism. It speeds up the rate at which certain changes take place in the body. Enzymes are essential in order for living cells to function properly. It is difficult *not* to get enough magnesium in one's daily diet. It is found in nuts, cereals, seafoods, and green vegetables. Most people have no problem getting the 300 to 400 milligrams of magnesium recommended in the daily diet.

A lack of magnesium can occur, however. For example, alcoholics and children in poor countries sometimes develop a magnesium deficiency. In such cases, magnesium deficiency may cause a person to become easily upset or overly aggressive.

However, it is also possible to be exposed to too much magnesium. For example, inhaling magnesium powder can produce irritation of the throat and eyes, resulting in a fever. In large doses, magnesium can cause damage to muscles and nerves. It can eventually result in loss of feeling and paralysis (inability to move parts of the body).

Such conditions are rare. They are likely to occur only among people who have to work with magnesium metal on a regular basis.

Manganese

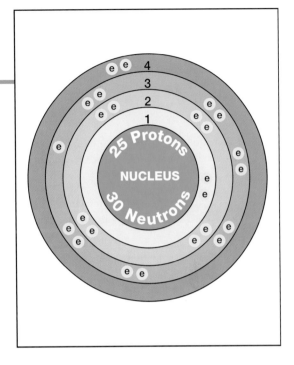

Overview

Manganese is a transition metal. The transition metals are the large block of elements in the middle of the periodic table. The periodic table is a chart that shows how chemical elements are related to each other. The transition metals make up Rows 4 through 7 in Groups 3 through 12 of the periodic table. Many of the best known and most widely used metals are in this group of elements.

It took chemists some time to discover the difference between manganese and **iron**. The two metals have very similar properties and often occur together in Earth's crust. The first person to clearly identify the differences between the two elements was Swedish mineralogist Johann Gottlieb Gahn (1745–1818) in 1774.

Manganese plays an interesting role in the U.S. economy. It is absolutely essential in the production of iron and steel. No element has been found that can replace manganese is such applications. The United States has essentially no manganese supplies of its own, so it depends on imports from other nations.

Key Facts

Symbol: Mn

Atomic Number: 25

Atomic Mass: 54.938045

Family: Group 7 (VIIB); transition metal

Pronunciation: MANG-guh-neez

WORDS TO KNOW

Allotropes: Forms of an element with different physical and chemical properties.

Alloy: A mixture of two or more metals with properties different from those of the individual metals.

Isotopes: Two or more forms of an element that differ from each other according to their mass number.

Machining: The bending, cutting, and shaping of a metal by mechanical means.

Periodic table: A chart that shows how the chemical elements are related to each other.

Radioactive isotope: An isotope that breaks apart and gives off some form of radiation.

Transition metal: An element in Groups 3 through 12 of the periodic table.

Discovery and Naming

One of the main ores of manganese is pyrolusite, which is made up primarily of the compound manganese dioxide (MnO_2). Early artists were familiar with pyrolusite. They used the mineral to give glass a beautiful purple color. They also used the mineral to remove color from a glass. When glass is made, it often contains impurities that give the glass an unwanted color. The presence of iron, for example, can give glass a yellowish tint. Adding pyrolusite to yellowish glass removes the color. The purple tint of pyrolusite balances out the yellow color of the glass. The glass ends up being clear and colorless.

By the mid-1700s, chemists began to suspect that pyrolusite might contain a new element. Some authorities credit German chemist Ignatius Gottfried Kaim with isolating the element in 1770. However, Kaim's report was not read by many chemists and was quickly lost.

During this period, some of the most famous chemists in Europe were trying to analyze pyrolusite, but none of them was successful. The problem was solved in 1774 when Gahn developed a method for removing the new element from pyrolusite. He heated pyrolusite with charcoal (pure **carbon**). The carbon took **oxygen** away from manganese dioxide, leaving behind pure manganese:

$$MnO_2 + C \rightarrow CO_2 + Mn$$

The origin of manganese's name is a bit confusing. Early chemists associated the new element with a mineral called magnesia. That mineral got its name from a region in Greece. Magnesia does not contain manganese, but the name stuck.

Physical Properties

Manganese is a steel-gray, hard, shiny, brittle metal. It is so brittle, in fact, that it cannot be machined in its pure form. Machining refers to the bending, cutting, and shaping of a metal by mechanical means. The melting point of manganese is 2,273°F (1,245°C) and its boiling point is about 3,800°F (2,100°C). Its density is 7.47 grams per cubic centimeter.

Manganese exists in four allotropic forms. Allotropes are forms of an element with different physical and chemical properties. The element changes from one form to another as the temperature rises. The form that exists from room temperature up to about 1,300°F (700°C) is the most common form.

Chemical Properties

Manganese is a moderately active metal. It combines slowly with oxygen in the air to form manganese dioxide (MnO_2). At higher temperatures, it reacts more rapidly. It may even burn, giving off a bright white light. Manganese reacts slowly with cold water, but more rapidly with hot water or steam. It dissolves in most acids with the release of **hydrogen** gas. It also combines with **fluorine** and **chlorine** to make manganese difluoride (MnF_2) and manganese dichloride ($MnCl_2$).

Occurrence in Nature

Manganese never occurs as a pure element in nature. It always combines with oxygen or other elements. The most common ores of manganese are pyrolusite, manganite, psilomelane, and rhodochrosite. Manganese is also found mixed with iron ores. In 2008, the largest producers of manganese ore in the world were South Africa, China, Australia, Gabon, and Brazil.

Manganese also occurs abundantly on the ocean floor in the form of nodules. These nodules are fairly large lumps of metallic ores. They usually contain **cobalt**, **nickel**, **copper**, and iron, as well as manganese. Scientists estimate that up to 1.5 trillion metric tons of manganese nodules may lie on the floors of the world's oceans and large lakes. Currently, there is no profitable method for retrieving these ores.

Rhodochrosite, which derives its name from the Greek words meaning rose-colored, is one of the most common ores of manganese. It is noted for its striking color. IMAGE COPYRIGHT 2009, MIAMIA. USED UNDER LICENSE FROM SHUTTERSTOCK.COM.

Manganese is the 12th most abundant element in Earth's crust. Its abundance is estimated to be 0.085 to 0.10 percent. That makes it about as abundant as fluorine or **phosphorus**.

Isotopes

Only one naturally occurring isotope of manganese exists: manganese-22. Isotopes are two or more forms of an element. Isotopes differ from each other according to their mass number. The number written to the right of the element's name is the mass number. The mass number represents the number of protons plus neutrons in the nucleus of an atom of the element. The number of protons determines the element, but the number of neutrons in the atom of any one element can vary. Each variation is an isotope.

Twenty-four radioactive isotopes of manganese are known also. A radioactive isotope is one that breaks apart and gives off some form of radiation. Radioactive isotopes are produced when very small particles are fired at atoms. These particles stick in the atoms and make them radioactive.

None of the radioactive isotopes of manganese has any important commercial uses.

Men and machine lay railroad tracks. A common alloy of manganese, ferromanganese, is contained in the steel used to produce railroad tracks. GALE GROUP.

Extraction

The usual method for producing pure manganese is to heat manganese dioxide (MnO_2) with carbon or **aluminum**. These elements remove the oxygen and leave pure manganese metal:

$$MnO_2 + C \rightarrow Mn + CO_2$$

Uses

Up to 90 percent of all manganese produced is made into steel alloys. An alloy is made by melting and mixing two or more metals. The mixture has properties different from those of the individual metals. The addition of manganese to steel makes the final product hard, as well as resistant to corrosion (rusting) and mechanical shock.

The most common alloy of manganese is ferromanganese. This alloy contains about 48 percent manganese combined with iron and carbon. Ferromanganese is the starting material for making a very large variety of steel products, including tools, heavy-duty machinery, railroad tracks, bank vaults, construction components, and automotive parts.

Another common alloy of manganese is silicomanganese. It contains manganese, **silicon**, and carbon in addition to iron. It is used for structural components and in springs.

Manganese is also used to make alloys with metals other than iron or steel. For example, the alloy known as manganin is 84 percent copper, 12 percent manganese, and 4 percent nickel. Manganin is used in electrical instruments.

Compounds

Less than 10 percent of all the manganese used in the United States goes to the production of manganese compounds. Perhaps the most important commercial use of these compounds is manganese dioxide (MnO_2). Manganese dioxide is used to make dry-cell batteries. These batteries are used in electronic equipment, flashlights, and pagers. Dry cell batteries hold a black pasty substance containing manganese dioxide. The use of manganese dioxide in a dry cell prevents hydrogen gas from collecting in the battery as electricity is produced.

Another manganese compound, manganous chloride ($MnCl_2$), is an additive in animal food for cows, horses, goats, and other domestic animals. Fertilizers also contain manganous chloride so that plants get all the manganese they need.

Finally, small amounts of manganese compounds are used as coloring agents in bricks, textiles, paints, inks, glass, and ceramics. Manganese compounds can be found in shades of pink, rose, red, yellow, green, purple, and brown.

Health Effects

Manganese is one of the chemical elements that has both positive and negative effects on living organisms. A very small amount of the element is needed to maintain good health in plants and animals. The manganese is used by enzymes in an organism. An enzyme is a substance that makes chemical reactions occur more quickly in cells. Enzymes are necessary to keep any cell operating properly. If manganese is missing from the diet, enzymes do not operate efficiently. Cells begin to die, and the organism becomes ill.

Fortunately, the amount of manganese needed by organisms is very small. It is not necessary to take extra manganese to meet the needs of cells.

In fact, an excess of manganese can create health problems. These problems include weakness, sleepiness, tiredness, emotional disturbances, and even paralysis. The only way to receive such a large dose is in a factory or mine. Workers may inhale manganese dust in the air.

Mercury

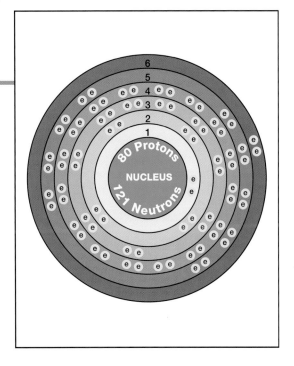

Overview

Mercury is a transition metal. A transition metal is one of the elements found between Groups 2 (IIA) and 13 (IIIA) in the periodic table. The periodic table is a chart that shows how chemical elements are related to one another. Mercury has long been known as quicksilver, because it is a silver liquid. The chemical symbol also reflects this property. The symbol, Hg, comes from the Latin term *hydrargyrum,* meaning "watery silver."

Mercury has been known for thousands of years. In many cultures, people learned to make mercury metal from its most important ore, cinnabar. When heated, cinnabar releases mercury as a vapor (gas). The vapor is cooled and captured as liquid mercury.

Some mercury compounds are known to be poisonous. For example, mercuric chloride (corrosive sublimate) was often used to kill pests and, sometimes, people. However, some mercury compounds have been used as medicines. For instance, mercurous chloride (calomel) was long used as a cure for skin rashes. Since the 1960s, the dangers of mercury have become better known. As a result, mercury use is now being phased out.

Key Facts

Symbol: Hg

Atomic Number: 80

Atomic Mass: 200.59

Family: Group 12 (IIB); transition metal

Pronunciation: MER-kyuh-ree

WORDS TO KNOW

Amalgam: A combination of mercury with at least one other metal.

Distillation: A process by which two or more liquids can be separated from each other by heating them to their boiling points.

Isotopes: Two or more forms of an element that differ from each other according to their mass number.

Periodic table: A chart that shows how the chemical elements are related to each other.

Phosphor: A material that gives off light when struck by electrons.

Radioactive isotope: An isotope that breaks apart and gives off some form of radiation.

Surface tension: A property of liquids that makes them act like they are covered with a skin.

Toxic: Poisonous.

Transition metal: An element in Groups 3 through 12 of the periodic table.

Discovery and Naming

The oldest sample of mercury dates to about the 15th or 16th century BCE. It was found in an Egyptian tomb at Kurna, stored in a small glass container.

Mercury and cinnabar are both mentioned in ancient manuscripts. The Chinese, Hindus, Egyptians, Greeks, and Romans all recorded information about the element and its ore. Greek philosopher Theophrastus (372–287 BCE), for example, described a method for preparing mercury. Cinnabar was rubbed together with vinegar in a clay dish. Theophrastus wrote that the cinnabar had been found in **silver** mines. When the metal was first made, he said, people thought it might contain gold. They were misled by the metal's shiny appearance. They soon realized, however, that it was quite different from **gold**.

Many reports on mercury told of its poisonous effects. Slaves who worked in Roman mercury mines, for example, often died of exposure to mercury. Strangely enough, trees and plants around these mines were not affected. Mercury was sometimes very dangerous and sometimes quite safe. People even drank from streams that ran through mercury mines. Scientists now know that mercury's effects depend on the form in which it occurs.

Mercury amalgams have also been around for a long time. An amalgam is a combination of mercury with at least one other metal. Amalgams are formed when a metal, such as silver, dissolves in mercury. The process is similar to dissolving salt in water. Amalgamation is used in mining to remove silver from ore. The silver dissolves in the mercury and a silver amalgam is formed. Heating the amalgam releases the silver. This method was used by miners as early as the 16th century.

Physical Properties

Mercury is the only liquid metal. In fact, there is only one other liquid element, **bromine**. Bromine is a non-metal. Mercury can be frozen (changed into a solid) at a temperature of −37.93°F (−38.85°C). It can be changed into a gas ("boiled") at 690.1°F (365.6°C). Its density is 13.59 grams per cubic centimeter, nearly 14 times the density of water.

Mercury has two physical properties of special interest. First, it has very high surface tension. Surface tension is a property of liquids that make them act like they are covered with a skin. For example, some water bugs are able to walk on the surface of water. With care, one can float a needle on the surface of water. These incidents are possible because of water's surface tension.

Mercury is also a very good conductor of electricity. This property is used in a number of practical devices. One such device is a mercury switch, such as the kind that turns lights on and off. A small amount of mercury can be placed into a tiny glass capsule. The capsule can be made

Droplets of mercury, the only liquid metal. IMAGE COPYRIGHT 2009, ANDRAZ CERAR. USED UNDER LICENSE FROM SHUTTERSTOCK.COM.

to tip back and forth. As it tips, the mercury flows from one end to the other. At one end of the capsule, the mercury may allow an electric current to flow through a circuit. At the other end, no mercury is present, so no current can flow. Mercury switches are easy to make and very efficient.

Chemical Properties

Mercury is moderately active. It does not react with **oxygen** in the air very readily. It reacts with some acids when they are hot, but not with most cold acids.

Occurrence in Nature

The abundance of mercury in Earth's crust is estimated to be about 0.5 parts per million. That makes it one of the 20 least common elements. It very rarely occurs as an element. Instead, it is usually found as a compound. Its most common ore is cinnabar, or mercuric sulfide (HgS). Cinnabar usually occurs as a dark red powder. It is often called by the common name of vermillion or Chinese vermillion.

According to the U.S. Geological Survey (USGS), as of 2008, the world's primary mercury resources can be found in China, Krygyzstan, Russia, Slovenia, Spain, and Ukraine. In the United States, mercury is found in states such as Alaska, Arkansas, California, Nevada, and Texas.

Isotopes

Seven naturally occurring isotopes of mercury are known. They are mercury-196, mercury-198, mercury-199, mercury-200, mercury-201, mercury-202, and mercury-204. Isotopes are two or more forms of an element. Isotopes differ from each other according to their mass number. The number written to the right of the element's name is the mass number. The mass number represents the number of protons plus neutrons in the nucleus of an atom of the element. The number of protons determines the element, but the number of neutrons in the atom of any one element can vary. Each variation is an isotope.

Thirty-six radioactive isotopes of mercury are known also. A radioactive isotope is one that breaks apart and gives off some form of radiation. Radioactive isotopes are produced when very small particles are fired at atoms. These particles stick in the atoms and make them radioactive.

Two radioactive isotopes of mercury are used in medicine: mercury-197 and mercury-203. Both isotopes are used to study the brain and the kidneys. The isotopes are injected into the body where they travel to the brain and the kidneys. Inside these two organs, the isotopes give off radiation that is detected by instruments held above the body. The pattern of radiation provides information about how well the brain and kidneys are functioning.

Extraction

Mercury is still prepared as it was hundreds of years ago. Cinnabar is heated in air. The compound breaks down to give mercury metal:

$$HgS \longrightarrow \text{heated} \rightarrow Hg + S$$

The mercury metal is then purified by distillation. Distillation is the process of heating two or more liquids to their boiling points. Different liquids boil at different temperatures. The liquid that is wanted (such as mercury) can be collected at *its* boiling point. Mercury that is more than 99 percent pure can be collected by distillation.

Uses

Among the most important uses of mercury is in the preparation of chlorine. Chlorine is produced by passing an electric current through sodium chloride:

$$2NaCl \longrightarrow \text{electric current} \rightarrow 2Na + Cl_2$$

There is a problem with using this method, however. Sodium (Na) is a very reactive metal. If any water is present, the sodium will react violently with the water. This reaction makes the production of chlorine much more difficult.

In 1892, two English chemists developed a method for solving this problem. They made a container with a layer of mercury on the bottom. As sodium is produced by the electric current, it dissolves in the mercury, forming an amalgam. The sodium is unable to react with water. For many years, the "mercury cell" invented in 1892 was a very popular method for producing chlorine.

But today, companies are looking for other ways to make chlorine. They are worried about the harmful effects of mercury. They are also concerned that mercury can get into the environment and harm humans, other animals, and plants.

Another of the most important uses of mercury in the United States is in switches and other electrical applications. Again, there are increasing concerns about the health effects of mercury. Many companies are switching to electronic switches.

One application in which concerns about mercury have had little effect is fluorescent lamps. A fluorescent lamp contains mercury vapor (gas). When the lamp is turned on, an electric current passes through the mercury vapor, causing it to give off invisible radiation. The radiation strikes the inside of the glass tube, whose walls are coated with a phosphor. A phosphor is a material that gives off visible light when struck by electrons. The tube glows as the radiation strikes the phosphor.

Lamp manufacturers have reduced the amount of mercury in fluorescent lamps by about 60 percent. They developed ways to make the lamps work just as well with less mercury. However, mercury lamps are much more popular. Each lamp now contains much less mercury. But there are many more lamps than ever before.

With fluorescent lights, when an electric current passes through mercury vapors, the resulting invisible radiation strikes phosphors. The phosphors then give off visible light. IMAGE COPYRIGHT 2009, STOCKSNAPP. USED UNDER LICENSE FROM SHUTTERSTOCK.COM.

For a time, mercury batteries were quite popular. In the early 1980s, more than 1,000 tons of mercury a year were used to make mercury batteries. These batteries create a special environmental problem, however. People tend to just throw them away when they no longer work. The cases split open easily, releasing mercury into the environment. As a result, much less mercury is now being used to make such batteries. In fact, the use of mercury for making batteries has been discontinued in the United States.

Mercury is also used in dental applications, measuring instruments (such as mercury thermometers and barometers), and coatings for mirrors.

Compounds

The use of mercury compounds is also decreasing because of health concerns. A few of the compounds still in use follow. Notice that two different endings are used for mercury compounds. Those that end in *-ous* have less mercury than those that end in *-ic*.

- mercuric arsenate ($HgHAsO_4$): water-proofing paints
- mercuric benzoate ($Hg(C_7G_5O_2)_2$): medicine; used to treat syphilis
- mercuric chloride, or mercury bichloride, or corrosive sublimate ($HgCl_2$): disinfectant, tanning of leather, spray for potato seedlings (to protect from disease), insecticide, preservation of wood, embalming fluid, textile printing, and engraving
- mercuric cyanide ($Hg(CN)_2$): germicidal soaps (soaps that kill germs), photography
- mercuric oxide (HgO): red or yellow pigment in paints, disinfectant, fungicide (to kill fungi), perfumes and cosmetics
- mercuric sulfide (HgS): red or black pigment in paints
- mercurous chloride, or calomel (Hg_2Cl_2): fungicide, maggot control in agriculture, fireworks
- mercurous chromate (Hg_2CrO_4): green pigment in paints
- mercurous iodide (Hg_2I_2): kills bacteria on the skin

"Mad as a Hatter"

Back in the 1800s, most of the negative effects of mercury and its compounds were not yet known. Hatmakers of that time commonly used a mercury compound in their craft. It was used to treat the felt and beaver fur that lined the hats. Eventually, exposure to the mercury began to cause changes in the hatmakers' bodies. Their personalities and behavior became erratic.

Recognizing the bizarre personalities of many hatmakers, people often used the expression "mad as a hatter." In fact, author Lewis Carroll (1832–1898) created a character for *Alice's Adventures in Wonderland* that owes its origins to the symptoms of mercury poisoning: The Mad Hatter.

Health Effects

Mercury metal and most compounds of mercury are highly toxic. Interestingly enough, scientists became aware of this fact only quite recently. However, the toxicity of *some* mercury compounds has been known for many centuries. One form of mercury chloride known as calomel, for example, was sometimes used as a poison to kill people. It was also once used extensively to kill fungi and control maggots in agricultural crops.

But even as recently as the mid-20th century, there was relatively little concern about mercury metal and many mercury compounds. High school chemistry students often played with tiny droplets of mercury in the laboratory. They used mercury to coat pennies and other pieces of metal. Mercury was also widely used in dentistry. It was used to make amalgams, alloys of mercury with other metals, used to fill teeth. Most people even today are likely to have dental fillings that contain a small amount of mercury metal. The question of whether mercury is still safe to use in dental fillings is the source of considerable controversy. Some people say that so little mercury is lost from fillings that the metal presents no danger to people. Other people think that dentists should take no chances with this dangerous metal. They should stop using mercury fillings entirely.

Since the mid-20th century, chemists have learned a great deal more about the toxic effects of both mercury metal and most of its compounds. They now know that mercury itself enters the body very easily. Its vapors pass through the skin into the bloodstream. Its vapors can also be inhaled. And, of course, it can also be swallowed. In any of these cases, mercury gets into blood and then into cells. There it interferes with essential chemical reactions and can cause illness and death.

Sometimes, these effects occur over very long periods of time. People who work with mercury, for example, may take in small amounts of mercury over months or years. Health problems develop very slowly. These problems can include inflammation of the mouth and gums; loosening of the teeth; damage to the kidneys and muscles; shaking of the arms and legs; and depression, nervousness, and personality changes.

People can also be exposed to large doses of mercury over short periods of time. In such cases, even more serious health problems can arise. These include nausea, vomiting, diarrhea, stomach pain, damage to the kidneys, and death in only a week or so.

Tragic Effects of Mercury Poisoning In a tragic irony, a scientist who was helping to improve the environment died as a result of her efforts. In 1997, Dartmouth College chemistry professor Karen Wetterhahn died of mercury poisoning. Less than a year earlier, she had been experimenting with dimethyl mercury when she spilled a tiny amount on her hands. Dimethyl mercury is one of the most toxic of mercury compounds.

Wetterhahn was studying the effects that heavy metals (mercury, chromium, lead, and arsenic) have on living things. She was concerned about how these elements pollute the environment and cause disease in people.

In August 1996, as Wetterhahn was transferring some dimethyl mercury to a tube, the accident occurred. She was wearing latex gloves, but they were not adequate protection against the dangerous chemical. The mercury seeped into her skin. Wetterhahn did not begin to feel the effects of the exposure until six months later. She then started losing her balance, slurring her speech, and suffering vision and hearing loss. Tests showed her system had 80 times the lethal dose of mercury. Wetterhahn died of mercury poisoning on June 8, 1997.

Wetterhahn's death prompted some safety changes. Bright stickers on latex glove boxes warn against using the gloves with hazardous chemicals. Workshops were held to teach proper glove selection. The dangers of dimethyl mercury were stressed. And scientists were urged to use a less dangerous chemical than dimethyl mercury. Overall, Wetterhahn's death heightened awareness in the scientific community of potential laboratory dangers.

Molybdenum

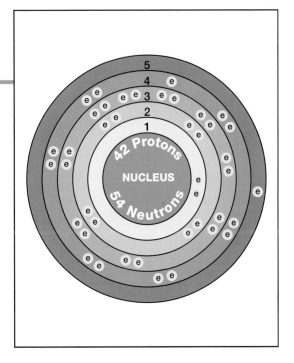

5
4
3
2
1

42 Protons

NUCLEUS

54 Neutrons

Overview

Molybdenum was one of the first metals to be discovered by a modern chemist. It was found in 1781 by Swedish chemist Peter Jacob Hjelm (1746–1813). Hjelm's work on the element was not published, however, until more than a century later.

Molybdenum is a transition metal, placing it in the center of the periodic table. The periodic table is a chart that shows how chemical elements are related to one another.

Molybdenum is a hard, silvery metal with a very high melting point. It is used primarily to make alloys with other metals. An alloy is a mixture of two or more metals. The mixture has properties different from those of the individual metals. The most common alloys of molybdenum are those with steel. Molybdenum improves the strength, toughness, resistance to wear and corrosion, and ability to harden steel.

Discovery and Naming

The most common ore of molybdenum is called molybdenite. Molybdenite contains a compound of molybdenum and **sulfur**, molybdenum disulfide (MoS_2). Molybdenum disulfide is a soft black powder that looks

Key Facts

Symbol: Mo

Atomic Number: 42

Atomic Mass: 95.96

Family: Group 6 (VIB); transition metal

Pronunciation: muh-LIB-duh-num

WORDS TO KNOW

Alloy: A mixture of two or more metals that has properties different from those of the individual metals.

Catalyst: A substance used to speed up a chemical reaction without undergoing any change itself.

Chemical reagent: A substance, such as an acid or an alkali, used to study other substances.

Isotopes: Two or more forms of an element that differ from each other according to their mass number.

Radioactive isotope: An isotope that breaks apart and gives off some form of radiation.

Radioactive tracer: An isotope whose movement in the body can be followed because of the radiation it gives off.

Trace element: An element that is needed in very small amounts for the proper growth of a plant or animal.

Transition metal: An element in Groups 3 through 12 of the periodic table.

---

like graphite. Graphite is pure **carbon**; it makes up the "lead" in ordinary pencils. In fact, earlier chemists thought that graphite and molybdenum disulfide were the same material.

The soft "squishy" character of molybdenum disulfide frustrated early researchers of the compound. Chemists often grind up a material before trying to dissolve it in acids or other liquids. But molybdenum disulfide cannot be ground up. The material just slides out of the way.

It was not until 1781 that Hjelm found a way to work with the compound. He discovered that it was very different from graphite. In fact, he found that it contained an entirely new element. The name chosen for the new element illustrates a further confusion. In Greek, the word for **lead** is *molybdos*. The name chosen for the new element, molybdenum, is actually the Greek word for lead!

Hjelm's work was known to his fellow chemists because of letters they had written to each other. But the report of his discovery was not actually printed for all chemists to read until 1890. Between 1791 and 1891, Hjelm's research was repeated by other chemists. They confirmed what he discovered, and he is recognized today as the discoverer of molybdenum.

Physical Properties

As a solid, molybdenum has a silvery-white metallic appearance. It more commonly occurs as a dark gray or black powder with a metallic luster.

Its melting point is about 4,700°F (about 2,610°C) and the boiling point is 8,600 to 10,000°F (4,800 to 5,560°C). Its density is 10.28 grams per cubic centimeter, more than 10 times the density of water.

Chemical Properties

Molybdenum does not dissolve in most common chemical reagents. A chemical reagent is a substance that takes part in a chemical reaction, such as an acid or an alkali. For example, molybdenum does not dissolve in hydrochloric acid, hydrofluoric acid, ammonia, sodium hydroxide, or dilute sulfuric acid. These chemicals are reagents often used to test how reactive a substance is. Molybdenum does dissolve in hot strong sulfuric or nitric acids, however. The metal does not react with **oxygen** at room temperatures, but does react with oxygen at high temperatures.

Occurrence in Nature

Molybdenum never occurs free in nature. Instead, it is always part of a compound. In addition to molybdenite, it occurs commonly as the mineral wulfenite ($PbMoO_4$). Its abundance in Earth's crust is estimated to be about 1 to 1.5 parts per million. That makes it about as common as **tungsten** and many of the rare earth (lanthanoid) elements. In 2008, the largest producers of molybdenum in the world included the United States, China, Chile, Peru, and Canada. In the United States, molybdenum ores were found primarily in Colorado, Idaho, Nevada, and New Mexico. According to the U.S. Geological Survey (USGS), the value of the molybdenum from U.S. mines was $4.5 billion that year.

Isotopes

Seven naturally occurring isotopes of molybdenum exist: molybdenum-92, molybdenum-94, molybdenum-95, molybdenum-96, molybdenum-97, molybdenum-98, and molybdenum-100. Isotopes are two or more forms of an element. Isotopes differ from each other according to their mass number. The number written to the right of the element's name is the mass number. The mass number represents the number of protons plus neutrons in the nucleus of an atom of the element. The number of protons determines the element, but the number of neutrons in the atom of any one element can vary. Each variation is an isotope.

None of the seven naturally occurring molybdenum isotopes is radioactive. However, 22 artificial radioactive isotopes have been produced.

A radioactive isotope is one that breaks apart and gives off some form of radiation. Radioactive isotopes are produced when very small particles are fired at atoms. These particles stick in the atoms and make them radioactive.

One radioactive isotope of molybdenum is commonly used in medicine, molybdenum-99m. (The "m" in this instance stands for "metastable," which means the isotope does not last very long.) This isotope is not used directly, however. Instead, it is used in hospitals to make another radioactive isotope, technetium-99m. This isotope of **technetium** (atomic number 43) is widely used as a tracer for diagnostic studies of the brain, liver, spleen, heart, and other organs and body systems.

A radioactive tracer is an isotope whose movement in the body can be followed because of the radiation it gives off. The radiation can be "traced" with special equipment held above the body. The pattern produced by the radiation allows a doctor to diagnose any unusual functioning (behavior) of the organ or body part.

Technetium-99m cannot be used for this purpose all by itself. It changes very quickly into a new isotope. Hospitals prepare molybdenum-99m first. This isotope can be stored for short periods of time. It slowly gives off radiation and changes into technetium-99m. The technetium-99m is captured as it is formed from molybdenum-99m and injected into the body for tracer studies. Because it is used to produce technetium-99m, the isotope molybdenum-99m is sometimes referred to as a "molybdenum cow."

Extraction

Pure molybdenum metal can be obtained from molybdenum trioxide (MoO_3) in a variety of ways. For example, hot hydrogen can be passed over the oxide to obtain the metal:

$$MoO_3 + 3H_2 \rightarrow 3H_2O + Mo$$

Uses

About 83 percent of the molybdenum used in the United States in 2008 was made into alloys of steel and iron. Some of these alloys, in turn, were used to make stainless and heat-resistant steel. A typical use is in airplane, spacecraft, and missile parts. Another important use of molybdenum alloys is in the production of specialized tools. Spark plugs, propeller shafts, rifle barrels, electrical equipment used at high temperatures, and boiler plates are all made of molybdenum steel.

Another important use of molybdenum is in catalysts. A catalyst is a substance used to speed up a chemical reaction. The catalyst does not undergo any change itself during the reaction. Molybdenum catalysts are used in a wide range of chemical operations, in the petroleum industry, and in the production of polymers and plastics.

Compounds

A number of molybdenum compounds are used in industry and research. Interestingly, molybdenum disulfide is still used as a lubricant, as it was more than 200 years ago. The slippery black powder looks and behaves much like graphite. Molybdenum sulfide is used in industrial operations to reduce the friction between sliding or rolling parts. It does not break down when heated or used for very long periods of time.

Other compounds of molybdenum are used as protective coatings in materials used at high temperatures; solders; as catalysts; as additives to animal feeds; and as pigments and dyes in glasses, ceramics, and enamels.

Health Effects

Molybdenum is relatively safe for humans and animals. No studies have shown it to be toxic. In fact, it is regarded as a necessary trace element for the growth of plants. A trace element is one that is needed in very small amounts for the proper growth of a plant or animal.

Neodymium

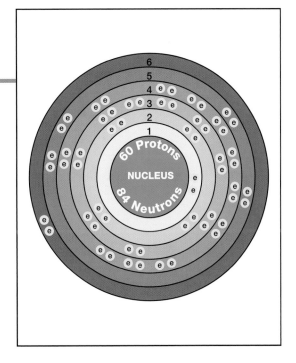

Overview

Neodymium was discovered in 1885 by Austrian chemist Carl Auer (Baron von Welsbach; 1858–1929). Auer found the new element in a mineral called didymia. Didymia, in turn, had been found in another complicated mineral known as ceria, originally found in Sweden in 1803. It took chemists nearly a century to completely analyze ceria. When they had done so, they found that it contained seven new elements. Neodymium was one of these elements.

Neodymium is in Row 6 of the periodic table. The periodic table is a chart that shows how chemical elements are related to each other. It is one of the rare earth elements. The term "rare earth" is inaccurate, however. These elements are not especially rare but were once difficult to separate from each other. These rare earth elements are also called the lanthanoids. That name comes from the third element in Row 6, **lanthanum**.

Neodymium has long been used in coloring glass and is now used in making lasers, very powerful magnets, and special alloys.

Discovery and Naming

During the late 1700s, two important mineral discoveries were made in Sweden. One was made just outside the town of Ytterby. The mineral

Key Facts

Symbol: Nd

Atomic Number: 60

Atomic Mass: 144.242

Family: Lanthanoid (rare earth metal)

Pronunciation: nee-oh-DIM-ee-um

WORDS TO KNOW

Isotopes: Two or more forms of an element that differ from each other according to their mass number.

Lanthanoids (rare earth elements): The elements in the periodic table with atomic numbers 57 through 71.

Laser: A device for producing very bright light of a single color.

Malleable: Capable of being hammered into thin sheets.

Periodic table: A chart that shows how the chemical elements are related to each other.

Reactive: Having a tendency to combine with other substances.

Tarnishing: Oxidizing; reacting with oxygen in the air.

found there, yttria, was eventually found to contain nine new elements. The second discovery was made near the town of Bastnas. That mineral, called cerite, was later found to contain seven new elements.

Cerite was thoroughly studied by Swedish chemist Carl Gustav Mosander (1797–1858). In 1839, Mosander was able to separate cerite into two parts, which he called **cerium** and lanthanum. Mosander believed he had found two new elements. Two years later, however, he learned that his lanthanum was not an element but a mixture of two parts. Mosander called these two new parts lanthanum and didymium. Mosander chose the name didymium because it means "twin." He said that didymium was like an identical twin to lanthanum. Chemists later confirmed that two of Mosander's discoveries were really new elements: cerium and lanthanum.

Mosander's didymium was not an element, however. In 1885, Auer found that didymium consisted of two simpler materials. The new elements were named neodymium and **praseodymium**. Auer chose the name neodymium because it means "new twin." Praseodymium, by comparison, means "green twin." Compounds of praseodymium are green.

Mosander, Auer, and other chemists of the time had only crude equipment with which to work. They never isolated any new element in a pure form. They found compounds of the element, usually a compound of the element and **oxygen**. The first pure samples of neodymium were not produced until 1925.

Physical Properties

Neodymium is a soft, malleable metal. Malleable means capable of being hammered into thin sheets. It can be cut and shaped fairly easily. It has a melting point of 1,875°F (1,024°C) and a boiling point of about 5,490°F (3,030°C). Neodymium has a density of 7.0 grams per cubic centimeter, seven times the density of water.

Chemical Properties

Neodymium is somewhat reactive. For example, it combines with oxygen in the air to form a yellowish coating. To protect it from tarnishing, the metal is usually stored in mineral oil and wrapped in plastic.

Neodymium shows typical properties of an active metal. For example, it reacts with water and acids to release **hydrogen** gas.

Occurrence in Nature

Neodymium is one of the most abundant of the rare earth elements. Its abundance in Earth's crust is thought to be about 12 to 24 parts per million. That places it about 27th among the chemical elements. It is slightly less abundant than **copper** and **zinc**.

The most common ores of neodymium are monazite and bastnasite. These ores are the most common source for all the rare earth elements.

Isotopes

Seven naturally occurring isotopes of neodymium are known. They are neodymium-142, neodymium-143, neodymium-144, neodymium-145, neodymium-146, neodymium-148, and neodymium-150. Five of the isotopes are stable and two, neodymium-144 and neodymium-150, are radioactive. Isotopes are two or more forms of an element. Isotopes differ from each other according to their mass number. The number written to the right of the element's name is the mass number. The mass number represents the number of protons plus neutrons in the nucleus of an atom of the element. The number of protons determines the element, but the number of neutrons in the atom of any one element can vary. Each variation is an isotope.

Twenty-nine radioactive isotopes of neodymium have also been produced artificially. A radioactive isotope is one that breaks apart and gives off some form of radiation. Radioactive isotopes are produced when very small particles are fired at atoms. These particles stick in the atoms and make them radioactive. None of neodymium's radioactive isotopes has any important use.

Predicting Volcanic Eruptions Rare earth elements have very special applications in scientific research. For example, consider a discovery made by scientists at the Lawrence Berkeley Laboratory (LBL) Center for Isotope Geochemistry in Berkeley, California. These scientists were trying to predict the size of a volcanic eruption. If they knew an eruption was going to occur, could they estimate how much lava it would produce?

Surprisingly, they found the answer in isotopes of neodymium. They discovered that large volcanic eruptions produced lava with one kind of isotope composition. Smaller eruptions produced lava with a different isotope composition.

So when a volcano starts producing lava, it can be studied for neodymium isotopes. From the composition of isotopes, scientists may be able to predict how big the coming eruption will be. The LBL scientists plan to use this information to warn residents of the intensity of the eruption.

Extraction

Neodymium occurs with other rare earth elements in monazite, bastnasite, and allanite. It must first be separated from these other elements.

It is then obtained in a pure form by reacting neodymium fluoride (NdF_3) with calcium:

$$3NdF_3 + 2Ca \rightarrow 2CaF_2 + 3Nd$$

Uses and Compounds

Neodymium and its compounds have a number of important uses. One is in a kind of laser known as a neodymium **yttrium aluminum** garnet (Nd:YAG) laser. A laser is a device for producing very bright and focused light of a single color. The Nd:YAG laser is used for treating bronchial cancer and certain eye disorders. The bronchi are air tubes that lead into the lungs.

Another important use of neodymium is in the manufacture of very strong magnets. The neodymium-**iron**-**boron** (NIB) magnet is one of the strongest magnets known. It is so strong it has to be handled with special care. Two NIB magnets can attract each other so strongly that they can smash into each other and shatter. An NIB magnet is inexpensive. Such magnets are used in stereo audio speakers.

Neodymium is also used in various optical (light) devices. For example, the General Electric Company makes a lightbulb called an "Enrich" bulb. The glass of the bulb contains a small amount of neodymium that filters out yellowish and greenish colors from the filament. The filament is the metal wire that is heated and gives off light. The light produced by the Enrich bulb is a very bright white light.

One of the oldest uses of neodymium is in coloring glass. The addition of a small amount of the element to glass gives it a greenish color. Some Tiffany lamp shades contain neodymium.

Neodymium is also used in various green (environmentally friendly) technologies, including hybrid cars and wind turbines.

Health Effects

Neodymium is regarded as moderately hazardous. Its compounds are known to irritate the eyes and skin. It should be handled with caution.

Neodymium-iron-boron magnets are used in audio speakers. IMAGE COPYRIGHT 2009, ANDI HAZELWOOD. USED UNDER LICENSE FROM SHUTTERSTOCK.COM.

Neon

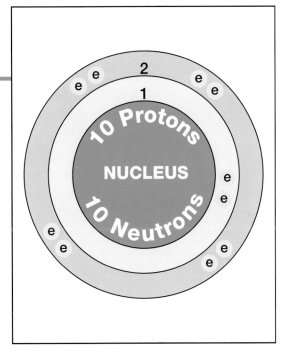

Overview

Neon is a member of the noble gas family. Other elements in this family include **helium**, **argon**, **krypton**, **xenon**, and **radon**. These gases are in Group 18 (VIIIA) of the periodic table. The periodic table is a chart that shows how chemical elements are related to each other.

The noble gases are sometimes called the inert gases. This name comes from the fact that these elements do not react very readily. In fact, compounds exist for only four noble gases—argon, krypton, radon, and xenon. Chemists have yet to prepare compounds of helium or neon.

Neon was discovered in 1898 by British chemists William Ramsay (1852–1916) and Morris Travers (1872–1961). It occurs naturally in the atmosphere, but only in very small amounts.

Neon has relatively few uses. The most familiar use is in neon lighting. Today, neon signs of every color, shape, and size exist. Neon signs are often filled with neon gas, but they may also contain other gases as well. The gas contained in the sign tube determines the color of light given off. The color given off by neon itself is reddish-orange.

Key Facts

Symbol: Ne

Atomic Number: 10

Atomic Mass: 20.1797

Family: Group 18 (VIIIA); noble gas

Pronunciation: NEE-on

WORDS TO KNOW

Isotopes: Two or more forms of an element that differ from each other according to their mass number.

Laser: A device for making very intense light of one very specific color that is intensified many times over.

Noble gas: An element in Group 18 (VIIIA) of the periodic table.

Periodic table: A chart that shows how the chemical elements are related to each other.

Radioactive isotope: An isotope that breaks apart and gives off some form of radiation.

Spectra: The lines produced when chemical elements are heated.

Spectroscopy: The process of analyzing light produced when an element is heated.

Discovery and Naming

It took humans centuries to understand air. At one time, philosophers thought air was an element. Many ancient Greek philosophers, for example, thought that the four basic elements were air, fire, water, and earth.

The first research to disprove that idea was done in the 1770s. In that decade, two new elements were discovered in air: **nitrogen** and **oxygen**. For some time, chemists were convinced that these two gases were the only ones present in air. That idea is easy to understand. Between them, nitrogen and oxygen make up more than 99 percent of air.

But over time, chemists became more skilled at making measurements. They recognized that something else was in air besides nitrogen and oxygen. That "something else" accounted for the remaining 1 percent that is not nitrogen or oxygen. In 1894, a third element was discovered in air: argon. Argon makes up about 0.934 percent of air. So, nitrogen, oxygen, and argon together make up about 99.966 percent of air.

But what was responsible for the remaining 0.034 percent of air? Chemists knew that other gases must be present in very small amounts. But what were those gases?

That question was answered between 1895 and 1900. Five more inert gases were discovered in air. One of those gases was neon.

Detecting gases in very small amounts was very difficult in the 1890s. Equipment was often not good enough to capture a tiny fraction of a milliliter of gas. But a new method, called spectroscopy, was developed that "sees" even small amounts of an element. Spectroscopy is the process of analyzing the light produced when an element is heated. The light pattern, or spectrum, produced is different for every element. The spectrum (plural: spectra) consists of a series of very specific colored lines.

In 1898, Ramsay and Travers were studying the minute amount of gas that remained after oxygen, nitrogen, and argon had been removed from air. They heated the sample of gas and studied the spectrum produced by it. Ramsay and Travers found spectral lines they had never seen before. They described their discovery:

> The blaze of crimson light from the tube told its own story, and it was a sight to dwell upon and never to forget. It was worth the struggle of the previous two years; and all the difficulties yet to be overcome before the research was finished. The *undiscovered gas* had come to light in a manner which was no less than dramatic. For the moment, the actual spectrum of the gas did not matter in the least, for nothing in the world gave a glow such as we had seen.

Ramsay's son was one of the first people to hear about the discovery. He wanted to name the new element *novum,* meaning "new." His father liked the idea, but suggested using the Greek word for "new," *neos.* Thus, the element was named neon.

Physical Properties

Neon is a colorless, odorless, tasteless gas. It changes from a gas to a liquid at –410.66°F (–245.92°C) and from a liquid to a solid at –415.5°F (–248.6°C). Its density is 0.89994 grams per liter. By comparison, the density of air is about 1.29 grams per liter.

Chemical Properties

Neon is chemically inactive. So far, it has been impossible to make neon react with any other element or compound.

Occurrence in Nature

The abundance of neon in normal air is 18.2 parts per million (0.0182 percent).

Isotopes

Three isotopes of neon exist in nature: neon-20, neon-21, and neon-22. Isotopes are two or more forms of an element. Isotopes differ from each

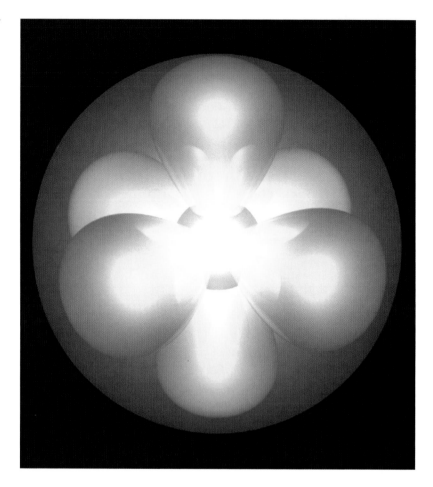

other according to their mass number. The number written to the right of the element's name is the mass number. The mass number represents the number of protons plus neutrons in the nucleus of an atom of the element. The number of protons determines the element, but the number of neutrons in the atom of any one element can vary. Each variation is an isotope.

Eleven radioactive isotopes of neon are known also. A radioactive isotope is one that breaks apart and gives off some form of radiation. Radioactive isotopes are produced when very small particles are fired at atoms. These particles stick in the atoms and make them radioactive.

None of the radioactive isotopes of neon has any commercial application.

Extraction

Neon can be obtained from air by fractional distillation. The first step in fractional distillation of air is to change a container of air to a liquid. The liquid air is then allowed to warm up. As the air warms, each element in air changes from a liquid back to a gas at a different temperature. The portion of air that changes back to a gas at −410.66°F (−245.92°C) is neon.

Uses

The best known use of neon gas is in neon lights. A neon light consists of a glass tube filled with neon or some other inert gas. An electric current is passed through the tube. The electric current causes neon atoms to break apart. After a fraction of a second, the parts recombine. When they recombine, they give off light. The color of light produced is determined by the gas used to fill the tube.

Neon lighting was invented by French chemist Georges Claude (1870–1960). Claude displayed his first neon sign at the Paris Exposition of 1910. He sold the first neon advertising sign to a Paris barber two years later.

By the 1920s, neon lighting had become popular in many parts of the world. Neon lights were fairly inexpensive, lasted a long time, and were very attractive. Probably the most spectacular collection of neon lighting is in Las Vegas, Nevada. Hotels, night clubs, and restaurants

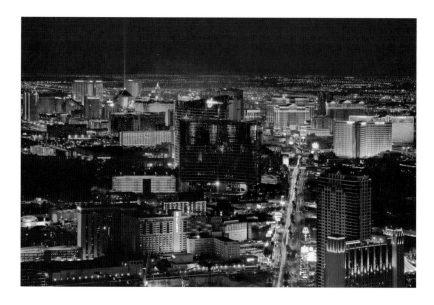

The glow of neon lights in Las Vegas, Nevada. IMAGE COPYRIGHT 2009, ANDY Z. USED UNDER LICENSE FROM SHUTTERSTOCK.COM.

seem to try to outdo everyone else in having the biggest and brightest neon sign.

Neon lighting is now used for many other purposes. For example, neon tubes are part of instruments used to detect electric currents. Neon is also used in the manufacture of lasers. A laser is a device for producing very bright light of a single color. Lasers now have many uses in industry and medicine. They are very efficient at cutting metal and plastic. They can also be used to do very precise kinds of surgery.

Compounds

There are no compounds of neon.

Health Effects

There are no known health effects of neon.

Neptunium

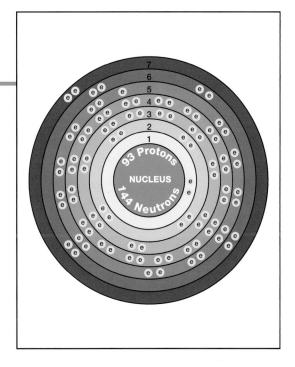

Overview

Neptunium lies in Row 7 of the periodic table. The periodic table is a chart that shows how chemical elements are related to one another. Neptunium is the first transuranium element. The term transuranium means "beyond **uranium**." Any element with an atomic number greater than 92 (uranium's atomic number) is called a transuranium element. Actinoid elements are found in Row 7. This name comes from element 89, **actinium**.

Scientists have now found more than 20 isotopes of neptunium. Neptunium was once a very rare element, but it can now be produced somewhat easily in a nuclear reactor. A nuclear reactor is a device in which nuclear fission reactions occur. Nuclear fission is the process of splitting atoms when neutrons collide with atoms of uranium or plutonium. These collisions produce new elements. Neptunium is used commercially only in specialized detection devices.

Discovery and Naming

The discovery of neptunium in 1940 represented an important breakthrough in the study of chemical elements. Scientists had known for

Key Facts

Symbol: Np

Atomic Number: 93

Atomic Mass: [237]

Family: Actinoid; transuranium element

Pronunciation: nep-TOO-nee-um

WORDS TO KNOW

Actinoid family: Elements with atomic numbers 89 through 103.

Half life: The time it takes for half of a sample of a radioactive element to break down.

Isotopes: Two or more forms of an element that differ from each other according to their mass number.

Nuclear fission: The process that occurs when neutrons collide with atoms of

uranium or plutonium, causing them to break apart.

Nuclear reactor: A device in which nuclear fission reactions occur.

Radioactive isotope: An isotope that breaks apart and gives off some form of radiation.

Transuranium (actinoid) element: An element with an atomic number greater than 92.

nearly a decade about an unusual kind of reaction. When an element is bombarded with neutrons, it sometimes changes into a new element. That new element has an atomic number one greater than the original element. For example, bombarding **copper** (atomic number 29) with neutrons may result in the production of **zinc** (atomic number 30). Bombarding **sodium** (atomic number 11) with neutrons may result in **magnesium** (atomic number 12).

One reason this discovery fascinated scientists was the possibility of bombarding uranium (atomic number 92) with neutrons. In the 1930s, uranium was the heaviest element known. It was the last element in the periodic table. But a "neutron change" like those described earlier would produce an element with atomic number 93. No one had ever heard of an element with atomic number 93!

In 1940, a pair of physicists at the University of California, Berkeley, were studying this problem. Edwin M. McMillan (1907–1991) and Philip H. Abelson (1913–2004) reported finding evidence of element number 93. They suggested naming it neptunium, in honor of the planet Neptune. (Uranium, the element before neptunium, had been named for the planet Uranus.)

Physical Properties

Neptunium is a silvery white metal with a melting point of 1,180°F (640°C) and a density of 20.45 grams per cubic centimeter, more than 20 times the density of water.

Chemical Properties

Neptunium is fairly reactive and forms some interesting compounds. Examples include neptunium dialuminide ($NpAl_2$) and neptunium beryllide ($NpBe_3$). These compounds are unusual because they consist of two metals joined to each other. Normally, two metals do not react with each other very easily. Neptunium also forms a number of more traditional compounds, such as neptunium dioxide (NpO_2), neptunium trifluoride (NpF_3), and neptunium nitrite ($NpNO_2$).

Occurrence in Nature

When neptunium was first discovered, scientists thought that it was an entirely artificial, or human-made, element. In 1942, very small amounts of the element were found in Earth's crust. The element can sometimes be found in ores containing uranium and other radioactive elements.

Isotopes

All isotopes of neptunium are radioactive. Isotopes are two or more forms of an element. Isotopes differ from each other according to their mass number. The number written to the right of the element's name is the mass number. The mass number represents the number of protons plus neutrons in the nucleus of an atom of the element. The number of protons determines the element, but the number of neutrons in the atom of any one element can vary. Each variation is an isotope.

One of the discoverers of neptunium, Edwin M. McMillan. LIBRARY OF CONGRESS.

A radioactive isotope is one that breaks apart and gives off some form of radiation. Radioactive isotopes are produced when very small particles are fired at atoms. These particles stick in the atoms and make them radioactive.

The longest lived isotope is neptunium-237. Its half life is 2,144,000 years. The half life of a radioactive element is the time it takes for half of a sample of the element to break down. Of a sample of neptunium-237,

Where Did They Go?: Disappearing Elements

Scientists think that Earth was formed about five billion years ago. What elements would a chemist have found on Earth in those days?

Part of the answer to that question is easy. Most of the elements found today were probably present five billion years ago. Those are the stable, or constant, elements. An untouched lump of gold in Earth's crust five billion years ago would still be a lump of gold today.

But that statement is not true for radioactive elements. Radioactive elements "fall apart" spontaneously. They break down and form new, simpler elements.

The rate at which radioactive elements break down differs from element to element, however. Some break down slowly, others break down quickly. Scientists measure the rate of breakdown in half lives. An element with a long half life breaks down very slowly. An element with a short half life breaks down quickly.

Uranium, for example, has three naturally occurring isotopes. Their half lives are 4.6 billion years, 700 million years, and 25 million years. If 100 metric tons of uranium were present when Earth was formed five billion years ago, about half of the first isotope would have broken down by now. About 50 metric tons of the element would remain. Scientists would have no trouble finding the element in Earth's crust.

But neptunium is a different story. Its longest lived isotope is neptunium-237, with a half life of about two million years. If 100 million tonnes of neptunium were present at Earth's beginning, only 50 million tons would be left after two million years. After another two million years (four million years altogether), only 25 million tons would be left. After another two million years (six million years altogether), only 12.5 million tons would be left.

Continue the mathematics. How much neptunium is left after 8 million years, 10 millions years, 12 million years, ...5 *billion* years? No need to do the calculations: Not very much neptunium at all would be left! Perhaps, too little to even find in Earth's crust.

So what does this example suggest about other transuranium elements, such as plutonium (number 94) and americium (number 95)? All of these elements have fairly short half lives. Of course, "fairly short" sometimes means "only" a few million years!

only half would remain after 2,144,000 years. The other half would have broken down to form new elements.

Neptunium-239 is the only isotope of neptunium to have practical uses. It is used in special instruments for detecting the presence of neutrons.

Extraction

Pure neptunium metal can be made by heating neptunium trifluoride (NpF_3) with hot barium or lithium metal:

Neptunium is used in nuclear reactors. Pictured here is Three Mile Island in Middletown, Pennsylvania, site of a partial meltdown in 1979. IMAGE COPYRIGHT 2009, DOBRESUM. USED UNDER LICENSE FROM SHUTTERSTOCK.COM.

$$3Ba + 2NpF_3 \rightarrow 3BaF_2 + 2Np$$

The metal can be purchased for legal uses.

Uses and Compounds

Neptunium and its compounds of neptunium have been made for research purposes. They are used in specialized detection devices and in nuclear reactors. Neither the element nor its compounds have any commercial uses.

Health Effects

Neptunium is a very hazardous material. The radiation it gives off can cause serious health problems for humans and other animals. It must be handled with great caution.

Radiation transfers large amounts of energy to cells and is quite penetrating. Cells that are damaged, but not killed, often reproduce out of control. This growth by functionally damaged cells forms tumors and causes related problems for organs and tissues.

Nickel

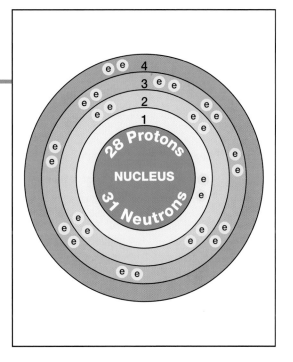

Overview

Nickel is the only element named after the devil. The name comes from the German word *Kupfernickel,* meaning "Old Nick's copper," a term used by German miners. They tried to remove copper from an ore that looked like copper ore, but they were unsuccessful. Instead of copper, they got slag, a useless mass of earthy material. The miners believed the devil ("Old Nick") was playing a trick on them. So they called the fake copper ore Old Nick's copper.

Since then, nickel has become a very valuable metal. The most common use is in the production of stainless steel, a strong material that does not rust easily. It is used in hundreds of industrial and consumer applications. Nickel is also used in the manufacture of many other alloys. An alloy is made by melting and mixing two or more metals. The mixture has properties different from those of the individual metals.

Nickel is classified as a transition metal. Transition metals are elements between Groups 2 (metals) and 13 (non-metals) in the periodic table. The periodic table is a chart that shows how chemical elements are related to one another. Nickel is closely related to **iron, cobalt, copper,** and **zinc**. These metals are close to nickel in the periodic table.

Key Facts

Symbol: Ni

Atomic Number: 28

Atomic Mass: 58.6934

Family: Group 10 (VIIIB); transition metal

Pronunciation: NI-kul

WORDS TO KNOW

Alloy: A mixture of two or more metals with properties different from those of the individual metals.

Ductile: Capable of being drawn into thin wires.

Electroplating: The process by which a thin layer of one metal is laid down on top of a second metal.

Isotopes: Two or more forms of an element that differ from each other according to their mass number.

Malleable: Capable of being hammered into thin sheets.

Metallurgy: The art and science of working with metals.

Nickel allergy: A health condition caused by exposure to nickel metal.

Periodic table: A chart that shows how the chemical elements are related to each other.

Radioactive isotope: An isotope that breaks apart and gives off some form of radiation.

Toxic: Poisonous.

Transition metal: An element in Groups 3 through 12 of the periodic table.

Discovery and Naming

The study of metals was difficult for early chemists. Many metals looked very similar. They also acted very much like each other chemically. Nickel was one of the metals about which there was much confusion.

Copper miners were confused about nickel and copper because they both occurred in ores with a green tint. But copper ores reacted differently to heat than did nickel ores. This confusion led to the choice for nickel's name.

But cobalt miners were confused too. Some ores of nickel also look like cobalt ores. But these ores did not react chemically in the same way either. Cobalt mine owners called the "misbehaving" ores of nickel "cobalt which had lost its soul."

Swedish mineralogist Axel Fredrik Cronstedt (1722–1765) was the first person to realize that nickel was a new element. In 1751, he was given a new mineral from a cobalt mine near the town of Hälsingland, Sweden. While Cronstedt thought the ore might contain cobalt or copper, his tests produced a surprising result. He found something in the mineral that did not act like cobalt, copper, or any other known element. Cronstedt announced that he had found a new element. He used a shortened version of Kupfernickel for the name of the new element. He called it nickel.

Physical Properties

Nickel is a silvery-white metal. It has the shiny surface common to most metals and is both ductile and malleable. Ductile means capable of being drawn into thin wires. Malleable means capable of being hammered into thin sheets. Its melting point is 2,831°F (1,555°C) and its boiling point is about 5,135°F (2,835°C). The density of nickel is 8.90 grams per cubic centimeter, nearly nine times the density of water.

Nickel is only one of three naturally occurring elements that is strongly magnetic. The other two are iron and cobalt. But nickel is less magnetic than either iron or cobalt.

Chemical Properties

Nickel is a relatively unreactive element. At room temperature, it does not combine with **oxygen** or water or dissolve in most acids. At higher temperatures, it becomes more active. For example, nickel reacts with oxygen to form nickel oxide (NiO):

$$2Ni + O_2 \text{ ---heat} \rightarrow 2NiO$$

It also reacts with steam to get nickel oxide and hydrogen gas:

$$Ni + H_2O \text{ ---heat} \rightarrow 2NiO + H_2$$

Occurrence in Nature

Nickel makes up about 0.01 to 0.02 percent of Earth's crust. It ranks about 22nd among the chemical elements in terms of abundance in Earth's crust. Nickel is thought to be much more abundant in Earth's core. In fact, many experts believe that the core consists almost entirely of iron and nickel.

One argument for this belief is the presence of nickel in meteorites. Meteorites are pieces of rock or metal from space that fall to Earth's surface. Meteorites often contain a high percentage of nickel.

The most common ores of nickel include pentlandite, pyrrhotite, and garnierite. The element also occurs as an impurity in ores of iron, copper, cobalt, and other metals.

Most of the nickel used in the United States is imported or obtained from recycling. As of 2008, the largest producers of nickel metal included Russia, Canada, Indonesia, Australia, and New Caledonia. One of the world's largest single deposits of nickel is located at Sudbury Basin,

Ontario, Canada. The deposit was discovered in 1883. It covers an area 17 miles (27 kilometers) wide and 37 miles (59 kilometers) long. Some experts believe the deposit was created when a meteorite struck the earth at Sudbury Basin.

Isotopes

There are five naturally occurring isotopes of nickel: nickel-58, nickel-60, nickel-61, nickel-62, and nickel-64. Isotopes are two or more forms of an element. Isotopes differ from each other according to their mass number. The number written to the right of the element's name is the mass number. The mass number represents the number of protons plus neutrons in the nucleus of an atom of the element. The number of protons determines the element, but the number of neutrons in the atom of any one element can vary. Each variation is an isotope.

Eighteen radioactive isotopes of nickel are known also. A radioactive isotope is one that breaks apart and gives off some form of radiation. Radioactive isotopes are produced when very small particles are fired at atoms. These particles stick in the atoms and make them radioactive.

One radioactive isotope of nickel, nickel-63, has limited use in industry. This isotope has two uses: for the detection of explosives, and in certain kinds of electronic devices, such as surge protectors. A surge protector is a device that protects sensitive electronic equipment like computers from sudden changes in the electric current flowing into them.

Extraction

The method used for making pure nickel metal is a common one in metallurgy. Metallurgy is the art and science of working with metals. Most nickel ores contain nickel sulfide (NiS). These ores are "roasted" (heated in air). Roasting converts the nickel sulfide to nickel oxide:

$$2NiS + 3O_2 \rightarrow 2NiO + 2SO_2$$

The nickel oxide is then treated with a chemical that will remove the oxygen from the nickel. For example:

$$2NiO + C \rightarrow CO_2 + 2Ni$$

A large amount of nickel is now recycled from scrap metal. Scrap metal comes from old cars, demolition of buildings, appliances like washing machines and stoves, and landfills. The task in recycling scrap metal is to find a way to separate the nickel from other metals in the scrap. This can be done by taking advantage of special properties of nickel. For example, a magnet will remove nickel from scrap, leaving copper behind.

Uses

One of the most important uses of nickel is in making alloys. About 86 percent of the primary nickel used in the United States in 2008 was used to make alloys. About half of that amount went into stainless steel. Stainless steel is common to household appliances (like coffee makers, toasters, and pots and pans), kitchen sink tops and stoves, and medical equipment (X-ray machines, for example). It is also used to make heavy machinery and large containers in which large-scale chemical reactions are carried out. Artists sometimes use stainless steel in sculpture because it does not rust easily. Stainless steel is important to the food and beverage, petroleum, chemical, pharmaceutical (drug), pulp and paper, and textile industries.

Nickel is also used to make the superalloys used in jet engine parts and gas turbines. Superalloys are made primarily of iron, cobalt, or nickel. They also include small amounts of other metals, such as **chromium**, **tungsten**, **aluminum**, and **titanium**. Superalloys are resistant to corrosion (rusting) and retain their properties at high temperatures.

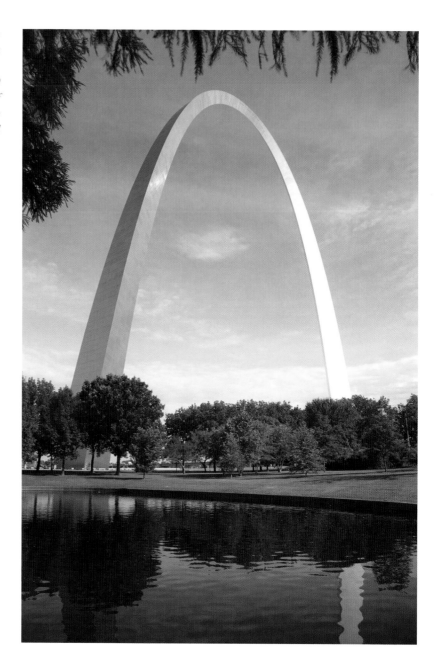

Nickel is also very popular in the manufacture of batteries. Nickel-cadmium (nicad) and nickel-metal hydride (NiMH) batteries are the most popular of these batteries. They are used in a great variety of appliances and electronics, including hand-held power tools, compact disc (CD)

players, portable DVD players, camcorders, cordless and cell telephones, scanner radios, and laptop computers.

Compounds

Some nickel compounds have important uses also. Many of these compounds are used in electroplating. Some are used to make alloys of nickel. Other nickel compounds are used as coloring agents. For example, the compound nickel dimethylglyoxime ($C_8H_{14}N_4NiO_4$) is used as a coloring agent in paints, cosmetics, and certain kinds of plastics.

Also, other nickel compounds have somewhat more unusual uses. For example, the compound nickel dibutyldithiocarbamate ($Ni[CS_2N(C_4H_9)_2]_2$) is used as an antioxidant in tires. The rubber in tires reacts with oxygen in the air. When it does so, the rubber gets hard and stiff. The tires begin to break down. An additive like nickel dibutyldithiocarbamate can reduce the rate at which this process occurs. The life of tires is extended.

Health Effects

Nickel can pose a health hazard to certain individuals. The most common health problem is called nickel allergy. Some people are more likely to develop nickel allergy than are others. People who are sensitive to nickel may develop a skin rash somewhat like poison ivy. The rash becomes

Explaining Electroplating

Nickel is also used in electroplating, a process by which a thin layer of one metal is laid down on top of a second metal. Here is how electroplating is done.

First, the nickel compound to be laid down is dissolved in water. The solution may be nickel chloride ($NiCl_2$), nickel nitrate ($Ni(NO_3)_2$), or some other nickel compound.

Second, a sheet of the metal to be electroplated is placed into the solution. Suppose the metal is steel. The steel sheet is suspended in the nickel chloride, nickel nitrate, or other nickel solution.

Third, an electric current is passed through the solution. The current causes nickel to come out of the solution. The nickel is then deposited on the surface of the steel. The longer the current runs, the more nickel is laid down. The thickness of the nickel layer can be controlled by the time the electric current runs through the solution.

Electroplating is used to make metal products with very specific qualities. Steel is strong but tends to corrode easily. Nickel does not corrode as fast as steel. A thin layer of nickel on top of steel protects the steel from corrosion.

itchy and may form watery blisters. Once a person gets nickel allergy, it remains with him or her forever.

Nickel is present in dozens of products. So it is easy for sensitive people to develop nickel allergy. Perhaps the most common cause of nickel allergy is body piercing. Some people have their ears pierced for earrings, while others have their lips, nose, or other body parts pierced. Inexpensive jewelry placed into these piercings is frequently made of stainless steel. Stainless steel contains nickel. The presence of nickel in a piercing can cause nickel allergy to develop.

Nickel can cause more serious health problems too. For example, people who are exposed to nickel fumes (dust and gas) breathe in nickel on a regular basis. Long term nickel exposure may cause serious health problems, including cancer.

Niobium

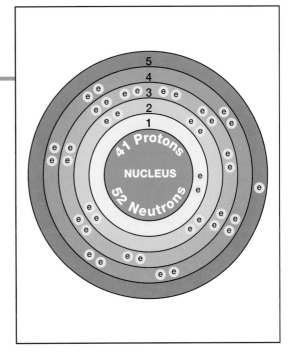

Overview

Niobium is a transition metal in Group 5 (VB) of the periodic table. The periodic table is a chart that shows how chemical elements relate to one another.

Niobium has a very interesting history. It was discovered by English chemist Charles Hatchett (1765–1847) in 1801. Hatchett found the element in a stone sent from North America. He named the element columbium. For years, scientists argued about the correct name for the element. Some still call the element columbium, although the official name is now niobium.

Niobium is used in many alloys. An alloy is made by melting and mixing two or more metals. The mixture has properties different from those of the individual metals. Niobium alloys are used in items that come into contact with the human body, such as rings for pierced ears, nose, and other body parts. Niobium is used in this kind of jewelry because it does not cause allergies or other problems.

Discovery and Naming

Historians today give credit for the discovery of niobium to Hatchett. The English chemist found the element in a "very heavy black stone, with

Key Facts

Symbol: Nb

Atomic Number: 41

Atomic Mass: 92.90638

Family: Group 5 (VB); transition metal

Pronunciation: nye-OH-bee-um

WORDS TO KNOW

Alloy: A mixture of two or more metals that has properties different from those of the individual metals.

Columbium: An alternative name for niobium.

Hypoallergenic: Not causing an allergic reaction.

Isotopes: Two or more forms of an element that differ from each other according to their mass number.

Radioactive isotope: An isotope that breaks apart and gives off some form of radiation.

Superconductivity: The tendency of an electric current to flow through a material without resistance.

golden streaks" that he found in the British Museum. The stone had been sent to England from the United States by John Winthrop (1681–1747). Winthrop was a member of the British Royal Society, one of the most important scientific societies in the world. (He was also the grandson of the first governor of Connecticut and the great-grandson of the first governor of Massachusetts.) The rock had been in the British Museum for nearly 70 years before anyone decided to analyze it. When Hatchett did so in 1801, he discovered a new element. He named it columbium, after the mineral columbite in which it is often found.

Not everyone agreed with Hatchett's discovery at first. Some chemists were convinced that columbium was identical to the element **tantalum**, which had been discovered earlier. The confusion is easy to understand. The two elements have very similar properties and are difficult to separate. Finally, in 1844, German chemist Heinrich Rose (1795–1864) showed that tantalum and columbium really *were* different from each other. Rose then suggested the name niobium for the new element. The name comes from Greek mythology. Niobe is the daughter of the god Tantalus, from whom the name tantalum comes.

Scientists debated for nearly a century over which name to use. In 1949, niobium was officially adopted. However, many metallurgists (scientists who work with metals) still use the name columbium for the element.

Physical Properties

Niobium is a shiny gray metal with a melting point of 4,474°F (2,468°C) and a boiling point of 8,901°F (4,927°C). Its density is 8.57 grams per cubic centimeter, more than eight times the density of water.

Chemical Properties

Niobium metal is resistant to attack by most common chemicals. It does not combine with **oxygen** or most other active elements except at high temperatures. It does not react with most strong acids unless they are hot and concentrated.

Occurrence in Nature

Niobium occurs primarily in two minerals, columbite and pyrochlore. The original name columbium was taken from the first of these minerals. Niobium always occurs with tantalum in these minerals. Separating the two elements is always the most difficult step in their preparation.

Columbite and pyrochlore are not mined in the United States. These two minerals are imported primarily from Brazil and Canada.

Scientists believe that niobium's abundance in Earth's crust is about 20 parts per million. That makes it about as abundant as **nitrogen** and **lithium**, and slightly more abundant than **lead**.

Isotopes

Only one naturally occurring isotope of niobium exists: niobium-93. Isotopes are two or more forms of an element. Isotopes differ from each other according to their mass number. The number written to the right of the element's name is the mass number. The mass number represents the number of protons plus neutrons in the nucleus of an atom of the element. The number of protons determines the element, but the number of neutrons in the atom of any one element can vary. Each variation is an isotope.

Forty-four radioactive isotopes of niobium are known also. A radioactive isotope is one that breaks apart and gives off some form of radiation. Radioactive isotopes are produced when very small particles are fired at atoms. These particles stick in the atoms and make them radioactive.

None of the radioactive isotopes of niobium have any practical application.

Extraction

The first step in preparing niobium metal is to separate its compounds from tantalum in an ore. The niobium compounds are then heated in air to change them to niobium oxide (Nb_2O_5). The niobium oxide is then heated with charcoal (**carbon**) to produce free metal:

$$2Nb_2O_5 + 5C \rightarrow 5CO_2 + 4Nb$$

Uses

Niobium is used primarily in making alloys. For example, the addition of niobium to steel greatly increases its strength. One use of such steel is in the construction of nuclear reactors. Nuclear reactors are devices in which the energy of nuclear reactions is converted to electricity. Niobium steel is used because it keeps its strength at the very high temperatures produced there.

The demand for niobium steel has increased. One reason is its increased use in airplanes and space vehicles. Some skateboards also include niobium steel components.

Another popular use of niobium alloys is in the making of jewelry. These alloys are lightweight and hypoallergenic. The term hypoallergenic means that they do not cause skin reactions. People who wear pierced earrings or similar forms of jewelry will not develop skin problems from niobium alloy jewelry.

Niobium alloys are also used in the construction of superconducting magnets. A superconducting material is one that has no resistance to an electric current. Once an electric current begins to flow in such a material, it continues to flow practically forever.

The most powerful magnets in the world are those made with superconducting materials. Scientists are continually looking to improve the strength of such magnets.

Compounds

Niobium diselenide ($NbSe_2$) is sometimes used as a lubricant at high temperatures. It does not break down at temperatures up to about 2372°F (1300°C). Niobium silicide ($NbSi_2$) is used as a refractory material. A refractory material is one that can withstand very high temperatures.

Niobium alloys are used in superconducting materials. They are used in applications ranging from magnetic resonance imaging (MRI) to levitation trains, like the prototype pictured here. Such trains use magnetic levitation and are capable of reaching high speeds. IMAGE COPYRIGHT 2009, MARTIN TRAJKOVSKI. USED UNDER LICENSE FROM SHUTTERSTOCK.COM.

Health Effects

Neither niobium nor its compounds are known to pose serious health effects for humans and animals.

Nitrogen

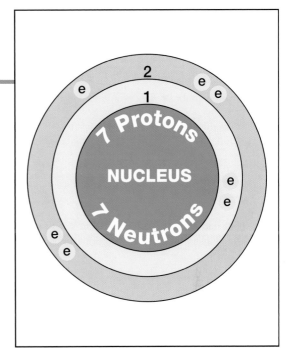

Overview

Nitrogen is the first member in Group 15 (VA) of the periodic table. The periodic table is a chart that shows how chemical elements are related to one another. Nitrogen is in a family group named after itself. Other elements in the nitrogen family are **phosphorus**, **arsenic**, **antimony**, and **bismuth**.

Nitrogen is one of the most interesting of all chemical elements. It is not a very active element. It combines with relatively few other elements at room temperature. Yet, the compounds of nitrogen are enormously important both in living organisms and in industrial applications. Five of the top 15 chemicals produced are compounds of nitrogen or the element itself. How does such an inactive element end up with so many important compounds?

Nitrogen makes up more than three-quarters of Earth's atmosphere. It is also found in a number of rocks and minerals in Earth's surface. It ranks about number 32 among the elements in terms of abundance in the earth.

Nitrogen was discovered by a number of chemists at about the same time, approximately 1772. But it was not until the early part of the 20th

Key Facts

Symbol: N

Atomic Number: 7

Atomic Mass: 14.0067

Family: Group 15 (VA); nitrogen

Pronunciation: NYE-tru-jun

WORDS TO KNOW

Anhydrous ammonia: Dry ammonia gas.

Catalyst: A substance used to speed up a chemical reaction without undergoing any change itself.

Isotopes: Two or more forms of an element that differ from each other according to their mass number.

Nitrogen fixation: The process of converting nitrogen as an element to a compound that contains nitrogen.

Periodic table: A chart that shows how the chemical elements are related to each other.

Proteins: Compounds that are vital to the building and growth of cells.

Radioactive isotope: An isotope that breaks apart and gives off some form of radiation.

Tracer: A radioactive isotope whose presence in a system can easily be detected.

century, when chemists learned how to make compounds of nitrogen, that these compounds became so widely used in a number of important applications.

By far the most notable use of nitrogen is in the production of ammonia (NH_3). Ammonia is used to make other compounds, such as ammonium sulfate (($NH_4)_2SO_4$), ammonium nitrate (NH_4NO_3), urea ($CO(NH_2)_2$), and nitric acid (HNO_3). These compounds are used primarily to make synthetic fertilizers. Both elemental nitrogen and nitrogen compounds have a number of important industrial uses.

Discovery and Naming

Gases were poorly understood by chemists until the late 1700s. What is air "made of?" That question is difficult to answer for a number of reasons. First, air cannot really be "seen." In fact, it took chemists many years to figure out how to capture air so that they could study it. Also, is ordinary "air" an element or a compound? For many centuries, philosophers said that air was an element. They could not imagine how anything as basic as air could be made of other materials.

Also, is ordinary air different from other kinds of "airs" seen in nature? For example, "air" sometimes comes bubbling out of the ground near oil wells. Today, scientists know that type of "air" is methane gas (CH_4). But early chemists were not sure how "oil air" differed from ordinary air.

Some important breakthroughs in the study of air occurred in the 1770s. The key was a simple experiment that science students still do today. The experiment begins with an empty bottle being turned upside down in a pan of water. The air in the bottle cannot get out.

If a burning candle is placed inside the bottle with the trapped air, the water rises just a bit. Why does this happen? Early chemists thought that a part of the air was used up as the candle burns. Today, they know that part of the air is **oxygen** gas. Another part of the air is left behind. That part does not disappear when the candle burns. This simple experiment shows that air is composed of (at least) two different elements: oxygen and something else.

Who's Who of Nitrogen Research One of the first people to discover the other unknown element was Scottish physician and chemist Daniel Rutherford (1749–1819). Rutherford carried out an experiment like the candle-in-a-bottle research just described.

Some of the greatest chemists of the time were working on this problem at the same time that Rutherford made his discovery. English chemist Henry Cavendish (1731–1810) probably discovered nitrogen before Rutherford did, but did not publish his findings. And in science, the first person to publish the results of an experiment usually gets credit for the work.

It seems likely that English chemist Joseph Priestley (1733–1804) and Swedish chemist Carl Wilhelm Scheele (1742–1786) also discovered nitrogen in the early 1770s.

Chemists debated about the name of this new element for some time. Antoine-Laurent Lavoisier (1743–1794), a French chemist who is often called the "father of modern chemistry," preferred the name azote meaning "without life." Lavoisier chose this name because nitrogen does not support breathing, the way oxygen does.

The modern name of nitrogen was first suggested in 1790 by French chemist Jean Antoine Claude Chaptal (1756–1832). This name made sense to chemists when they realized that the new gas was present in both nitric acid and nitrates. Thus, nitrogen means "nitrate and nitric acid" (*nitro-*) and "origin of" (*-gen*).

Physical Properties

Nitrogen is a colorless, odorless, tasteless gas with a density of 1.25046 grams per liter. By comparison, the density of air is about 1.29 grams per liter. Nitrogen changes from a gas into a liquid at a temperature of –320.42°F (–195.79°C). It changes from a liquid to a solid at a temperature of –346.02°F (–210.01°C). When it freezes, it becomes a white solid that looks like snow. Nitrogen is slightly soluble in water. About two liters of nitrogen can be dissolved in 100 liters of water.

Chemical Properties

At room temperature, nitrogen is a very inactive gas. It does not combine with oxygen, **hydrogen**, or most other elements. Nitrogen will combine with oxygen, however, in the presence of lightning or a spark. The electrical energy from either of those sources causes nitrogen and oxygen to form nitric oxide:

$$N_2 + O_2 \text{ —electrical energy} \rightarrow 2NO$$

Nitric oxide is more active than free nitrogen. For example, nitric oxide combines with oxygen and water in the atmosphere to make nitric acid. When it rains, nitric acid is carried to the earth. There it combines with metals in Earth's crust. Compounds known as nitrates and nitrites are formed.

Changing nitrogen as an element to nitrogen in compounds is called nitrogen fixation. The reaction between nitrogen and oxygen in the air when lightning strikes is an example of nitrogen fixation.

Certain bacteria have developed methods for fixing nitrogen. These bacteria live on the root hairs of plants. They take nitrogen out of air dissolved in the ground and convert it to compounds, such as nitrates. Those nitrates are used to make protein molecules, compounds vital to the building and growth of cells.

Plants, humans, and other animals do not have the ability to fix nitrogen. All living organisms on Earth depend on soil bacteria to carry out this process. Plants can grow because the bacteria fix nitrogen for them. They use the fixed nitrogen to make proteins. Humans and other animals can survive because they eat plants. They also depend on the soil bacteria that allow plants to make proteins. So all living creatures rely on soil bacteria to fix their nitrogen for them and, therefore, to survive.

Occurrence in Nature

Nitrogen is a fairly common element in Earth's crust. It occurs primarily as nitrates and nitrites. Nitrogen is by far the most important element in Earth's atmosphere. It makes up 78.084 percent of the atmosphere.

Isotopes

Two naturally occurring isotopes of nitrogen exist: nitrogen-14 and nitrogen-15. Isotopes are two or more forms of an element. Isotopes differ from each other according to their mass number. The number written to the right of the element's name is the mass number. The mass number represents the number of protons plus neutrons in the nucleus of an atom of the element. The number of protons determines the element, but the number of neutrons in the atom of any one element can vary. Each variation is an isotope.

Ten radioactive isotopes of nitrogen are known also. A radioactive isotope is one that breaks apart and gives off some form of radiation. Radioactive isotopes are produced when very small particles are fired at atoms. These particles stick in the atoms and make them radioactive.

None of the radioactive isotopes of nitrogen has any important commercial use. However, nitrogen-15 is used quite often in tracer studies. A tracer is a radioactive isotope whose presence in a system can be detected. Normally, tracer studies use radioactive isotopes. These isotopes give off radiation that can be detected with instruments. Nitrogen-15 is used for a different reason. A compound made with nitrogen-15 will weigh just a little bit more than one made with nitrogen-14. There are simple chemical methods for detecting whether a heavier compound or a lighter one is present in a system. Thus, nitrogen-15 can be used to trace the path of nitrogen through a system.

Extraction

Nitrogen is almost always made from liquid air. Liquid air is made by cooling normal atmospheric air to very low temperatures. As the temperature drops, the gases contained in air turn into liquids. At −297.33°F (−182.96°C), oxygen changes from a gas into a liquid. At −320.42°F (−195.79°C), nitrogen changes from a gas into a liquid. And so on. Eventually, all the gases in air can be made to liquefy (change into a liquid).

The reverse process also takes place. Suppose liquid air in a container warms up slowly. When its temperature reaches −320.42°F (−195.79°C), liquid nitrogen changes back to a gas. A container can be put into place to catch the nitrogen as it boils off the liquid air. When the temperature reaches −297.33°F (−182.96°C), oxygen changes from a liquid back to a gas. Another container can be put into place. The escaping oxygen can be collected. All of the gases in atmospheric air can be produced by this method.

Large amounts of nitrogen gas are produced in this way. In fact, nitrogen is second only to sulfuric acid in terms of production. In the early 2000s, more than 8.85 million metric tons (about 9.8 million short tons) of nitrogen gas were produced in the United States alone.

Uses

Nitrogen gas is used where an inert atmosphere is needed. An inert atmosphere is one that does not contain active elements. Ordinary air is not an inert atmosphere. It contains oxygen. Oxygen tends to react with other elements.

Suppose an ordinary lightbulb were filled with air. When the bulb is turned on, an electric current runs through the metal filament (wire) inside the bulb. The filament gets very hot, begins to glow, and gives off light.

But a hot metal wire will react quickly with oxygen in ordinary air. The metal combines with oxygen to form a compound of the metal. The metal compound will not conduct an electric current. The bulb will "burn out" very quickly.

An easy solution to that problem is to use nitrogen instead of ordinary air in the lightbulb. Nitrogen does not react with other elements very well, even when they get hot. The filament can get very hot, but the metal of which it is made will not combine with nitrogen gas. The nitrogen gas is an inert atmosphere for the bulb.

Another use for inert atmospheres is in protecting historic documents. Suppose the Declaration of Independence were simply left on top of a table for people to see. The paper and ink in the document would soon begin to react with oxygen in the air. They would both begin to decay. Before long, the document would begin to fall apart.

Instead, important documents like the Declaration of Independence are kept in air-tight containers filled with nitrogen gas. The documents are protected from oxygen and other gases in the air with which they might react. Although nitrogen gas provides an inert atmosphere for many purposes, other gases—especially argon—do an even better job.

Fairly simple methods are now available for changing nitrogen gas into liquid nitrogen. Liquid nitrogen is used to freeze other materials.

The temperature of the nitrogen has to be reduced to −320.42°F (−195.79°C) for this change to occur.

Today, it is possible to buy large containers of liquid nitrogen. The liquid nitrogen can be used, then, to freeze other materials. For example, foods can be frozen simply by dipping them into large vats of liquid nitrogen. Many frozen foods in a grocery store are usually produced this way. Liquid nitrogen can also be used to keep foods cold when they are being transported from one place to another.

Compounds

Nitrogen is the starting point for an important group of compounds. First, nitrogen is combined with hydrogen to make ammonia (NH_3). The production of ammonia is sometimes called industrial nitrogen fixation.

The formation of ammonia from nitrogen and hydrogen is very difficult to accomplish. The two elements do not easily combine. Finding a way to make nitrogen and hydrogen combine was one of the great scientific discoveries of the 20th century.

That discovery was made by German chemist Fritz Haber (1868–1934) in 1905. He found that nitrogen and hydrogen would combine

Nitrogen's Role in the Oklahoma City Bombing

On April 19, 1995, a bomb exploded at the Alfred P. Murrah Federal Building in Oklahoma City, Oklahoma. The devastating impact of the explosion destroyed the building within eight seconds. Each of the nine floors collapsed on top of one another. The bomb killed 168 people and injured hundreds more. The nation was stunned to learn that the attack had not been committed by international terrorists, but by a domestic terrorist, American Timothy McVeigh.

While searching the home of McVeigh's accomplice (a partner in a crime), Terry Nichols, investigators found a receipt for ammonium nitrate. This nitrogen compound is most commonly purchased by farmers. They use it as a fertilizer for their crops. But it can also be used as an explosive. The amount on the receipt was for 2,000 pounds. Since neither Nichols nor McVeigh was farming at the time the fertilizer was purchased, there was no real reason to purchase such a large amount of fertilizer.

Investigators determined that an ammonium nitrate bomb did, indeed, destroy the Murrah Building. The explosion resulted from a mixture of 4,800 pounds of ammonium nitrate and fuel oil from 20 plastic drums.

In June 1997, McVeigh was found guilty of all 11 charges against him, including eight counts of first-degree murder of federal agents. He was sentenced to death later that month and was put to death for his crime on June 11, 2001.

if they were heated to a very high temperature with a very high pressure. He also found that a catalyst was needed to make the reaction occur. The catalyst he used was **iron** metal, though other metals are sometimes used. A catalyst is a substance used to speed up a chemical reaction. The catalyst does not undergo any change itself during the reaction.

The significance of Haber's discovery was soon apparent. World War I began in 1914 in Europe. Very soon, Germany was no longer able to get nitrates from Chile. Nitrates were essential to the war effort for making explosives. But ships carrying nitrates from Chile to Germany were usually not able to get across the Atlantic Ocean.

However, nitrates can be made from ammonia using the Haber process. Germany was then able to make all the ammonia it needed. The ammonia was converted to nitrates for explosives. So a lack of nitrates from Chile did not stop the German war machine. Instead, Germany went on fighting for another four years against countries such as France and Great Britain, who were later joined by the United States.

Ammonia consistently ranks among the top 10 chemicals in the United States. The most important use of ammonia is in synthetic

The bomb used to destroy the federal building in Oklahoma City in 1995 was made of ammonium nitrate. The bomb killed 168 people, including more than a dozen children, while injuring 500 others.
AP IMAGES.

fertilizers. A synthetic fertilizer is a mixture of compounds used to make plants grow better. Most farmers use huge amounts of synthetic fertilizer every year to ensure large crops.

In 2008, about 89 percent of ammonia use in the United States was in fertilizer. This included anhydrous ammonia. Anhydrous means "without water." Anhydrous ammonia is simply ammonia gas. It is stored in large tanks. Farmers inject anhydrous ammonia directly into the ground to produce strong and healthy plants.

Ammonia is also found in many household cleaners, especially glass-cleaning and grease-cutting products.

The largest producer of ammonia in the world is China, which makes four times as much ammonia as the next leading producers (India and Russia). Other large producers are the United States, Trinidad and Tobago, Indonesia, Ukraine, Canada, Germany, Saudi Arabia, and Pakistan.

Ammonia can also be converted into other nitrogen compounds. For example, it can be combined with nitric acid (HNO_3) to form ammonium nitrate (NH_4NO_3). And it can be combined with sulfuric acid (H_2SO_4) to make ammonium sulfate (($NH_4)_2SO_4$):

$$NH_3 + HNO_3 \rightarrow NH_4NO_3$$

$$2NH_3 + H_2SO_4 \rightarrow (NH_4)_2SO_4$$

In the early 2000s, about 6.6 million short tons (6.0 million metric tons) of ammonium nitrate and about 2.9 million short tons (2.6 million metric tons) of ammonium sulfate were produced as fertilizers. These two compounds ranked number 14 and number 21 among chemicals produced in the United States.

Ammonium nitrate and ammonium sulfate both have other uses also. For example, ammonium nitrate is used to make explosives, fireworks, insecticides and herbicides (chemicals that kill insects and weeds), and rocket fuel. Ammonium sulfate is also used in water treatment systems, as a food additive, in the tanning of leather, in fireproofing materials, and as a food additive.

Yet another important compound of nitrogen is nitric acid (HNO_3). Nitric acid is made by reacting ammonia with oxygen:

$$NH_3 + 2O_2 \text{ — heat} \rightarrow HNO_3 + H_2O$$

Nitric acid usually ranks about number 13 among chemicals produced in the United States. The major use of nitric acid is to make ammonium nitrate as a synthetic fertilizer. Nitric acid is also used to make explosives, dyes, certain kinds of synthetic rubber and plastics, and in the preparation of metals.

Health Effects

Nitrogen is absolutely essential to all living organisms. It is an important part of all protein molecules. Proteins are the building material in all kinds of cells. They are also used for many other functions. For example, all living organisms use hormones to send chemical messages from one cell to another. Hormones are proteins.

Nitrogen is also used to make nucleic acids. Nucleic acids have many important functions in living organisms. They store the organism's genetic information, which is the set of instructions that tell every cell what its job in the organism is. It passes on that information from one generation to the next.

However, according to the U.S. Geological Survey (USGS) in 2009, recent studies indicate that various alpine and pristine lakes are being impacted by increasing levels of nitrogen emissions. Such emissions come from several sources: agriculture, energy production, and vehicle emissions. Scientists are studying the situation to learn how much the added nitrogen is impacting ecosystems, water quality, and other important issues.

Osmium

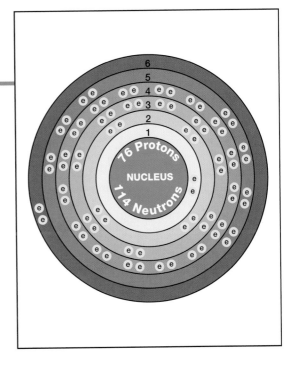

Overview

Osmium is an element in Group 8 (VIIIB) of the periodic table. The periodic table is a chart showing how chemical elements are related to one another. Osmium is also a member of the **platinum** family. This family consists of five other elements: **ruthenium**, **rhodium**, **palladium**, **iridium**, and platinum. These elements often occur together in Earth's crust. They also have similar physical and chemical properties, and they are used in alloys.

Osmium was discovered in 1804 by English chemist Smithson Tennant (1761–1815). Tennant found the new element in an ore of platinum.

Osmium is a very rare element and has few commercial uses. Osmium tetroxide (OsO_4) is more widely used, however, because it is so active.

Discovery and Naming

Platinum metal (atomic number 78) was known to chemists as early as 1741. Over the next 60 years, however, scientists discovered that the substance they knew as "platinum" was usually a mixture of substances.

Key Facts

Symbol: Os

Atomic Number: 76

Atomic Mass: 190.23

Family: Group 8 (VIIIB); transition metal; platinum group

Pronunciation: OZ-mee-um

WORDS TO KNOW

Alloy: An alloy is a mixture of two or more metals that has properties different from those of the individual metals.

Aqua regia: A mixture of hydrochloric and nitric acids that dissolves many materials that are not dissolved by either acid by itself.

Catalyst: A substance used to speed up a chemical reaction without undergoing any change itself.

Isotopes: Two or more forms of an element that differ from each other according to their mass number.

Periodic table: A chart that shows how the chemical elements are related to each other.

Radioactivity: Having a tendency to break apart and give off some form of radiation.

Toxic: Dangerous.

These substances proved to be new elements. Osmium was one of the new elements discovered in impure platinum.

In the early 1800s, Smithson Tennant was studying platinum. He found that a black powder remained when platinum was dissolved in aqua regia. Aqua regia is a mixture of hydrochloric and nitric acids. The term aqua regia means "royal water." It often dissolves materials that either acid by itself does not dissolve.

In 1804, Tennant announced that the black powder was actually a mixture of two new elements. He called them iridium and osmium. He suggested osmium's name because of the unusual smell of the compound he was working with, osmium tetroxide. Osmium comes from the Greek word *osme,* meaning "odor."

Physical Properties

Osmium is a bluish-white, shiny metal with a melting point of about 5,400°F (3,000°C) and a boiling point of about 9,900°F (5,500°C). Its density is 22.5 grams per cubic centimeter. These numbers are the highest of any platinum metal. They are also among the highest of all elements.

Osmium is unworkable as a metal. It cannot be melted and shaped like most metals. Because it is unworkable, it has very few practical uses.

Chemical Properties

Osmium is dissolved by acids or by aqua regia only after long periods of exposure to the liquids. When heated, the metal combines with **oxygen** to form osmium tetroxide (OsO_4). Osmium tetroxide is very toxic and the only important commercial compound of osmium.

Occurrence in Nature

Osmium is very rare. Its abundance is thought to be about 0.001 parts per million (one part per billion). That places the element among the half dozen least abundant elements in Earth's crust.

The most common ore of osmium is osmiridium. The element also occurs in all ores of platinum.

Isotopes

There are seven naturally occurring isotopes of osmium. The most abundant are osmium-192, osmium-190, and osmium-189. These three isotopes make up 41 percent, 26.4 percent, and 16.1 percent of natural osmium, respectively. Isotopes are two or more forms of an element. Isotopes differ from each other according to their mass number. The number written to the right of the element's name is the mass number. The mass number represents the number of protons plus neutrons in the nucleus of an atom of the element. The number of protons determines the element, but the number of neutrons in the atom of any one element can vary. Each variation is an isotope.

Two of the naturally occurring isotopes of osmium are radioactive: osmium-184 and osmium-186. A radioactive isotope is one that breaks apart and gives off some form of radiation. Radioactive isotopes are produced when very small particles are fired at atoms. These particles stick in the atoms and make them radioactive. In addition to the two naturally occurring radioactive isotopes of osmium, 33 radioactive isotopes have been prepared in the laboratory. None of these isotopes has any significant practical application.

Extraction

Osmium is obtained when platinum metal is extracted from its ores.

Uses

Osmium metal has few uses. It is sometimes added to platinum or iridium to make them harder. The osmium-platinum alloy is harder than pure platinum. Some alloys of osmium and platinum are used to make specialized laboratory equipment. Also, some of the best pen points are made of osmium-platinum alloys. An alloy is made by melting or mixing two or more metals.

Finely divided osmium metal is also used as a catalyst. A catalyst is a substance used to speed up a chemical reaction. The catalyst does not undergo any change itself during the reaction. The process for making ammonia from **hydrogen** and **nitrogen** sometimes uses osmium as a catalyst.

Compounds

Osmium tetroxide (OsO_4) is in demand for use as a catalyst for research purposes. The problem is that this compound of osmium is very dangerous to use. It is shipped in small glass containers called ampules. The ampules carry no labels, nor are they marked with ink. The label and ink would react violently with osmium tetroxide. Users are instructed to open and use an ampule containing osmium tetroxide with great care.

Health Effects

Some compounds of osmium are extremely dangerous. They irritate the respiratory passage (throat, lungs, etc.), the skin, and the eyes. They must be handled with extreme care. This caution is especially important for the most widely used compound of osmium, osmium tetroxide.

Oxygen

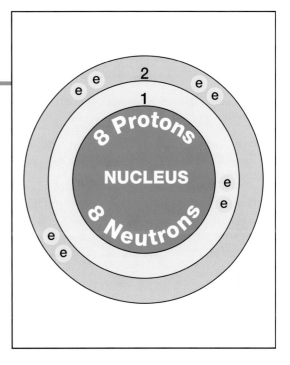

Overview

Oxygen is the first element in Group 16 (VIA) of the periodic table. The periodic table is a chart that shows how chemical elements are related to each other. The elements in Group 16 are said to belong to the chalcogen family. Other elements in this group include **sulfur**, **selenium**, **tellurium**, and **polonium**. The name chalcogen comes from the Greek word *chalkos,* meaning "ore." The first two members of the family, oxygen and sulfur, are found in most ores.

Oxygen is by far the most abundant element in Earth's crust. Nearly half of all the atoms in the earth are oxygen atoms. Oxygen also makes up about one-fifth of Earth's atmosphere. Nearly 90 percent of the weight of the oceans is due to oxygen. In addition, oxygen is thought to be the third most abundant element in the universe and in the solar system.

The discovery of oxygen is usually credited to Swedish chemist Carl Wilhelm Scheele (1742–1786) and English chemist Joseph Priestley (1733–1804). The two discovered oxygen at nearly the same time in 1774, working independently of each other.

Oxygen is necessary for the survival of all animal life on Earth. Animals breathe in oxygen and breathe out carbon dioxide. One important

Key Facts

Symbol: O

Atomic Number: 8

Atomic Mass: 15.9994

Family: Group 16 (VIA); chalcogen

Pronunciation: OK-si-jun

WORDS TO KNOW

Allotropes: Forms of an element with different physical and chemical properties.

Alloy: A mixture of two or more metals with properties different from those of the individual metals.

Isotopes: Two or more forms of an element that differ from each other according to their mass number.

Periodic table: A chart that shows how the chemical elements are related to each other.

Radioactive isotope: An isotope that breaks apart and gives off some form of radiation.

Rusting: A process by which a metal combines with oxygen.

use of oxygen is in medicine. People who have trouble breathing are given extra doses of oxygen. In many cases, this "extra oxygen" keeps people alive after they would otherwise have died.

But oxygen has many commercial uses also. The most important use is in the manufacture of metals. More than half of the oxygen produced in the United States is used for this purpose. Oxygen usually ranks among the top five chemicals produced in the United States each year. The gas is prepared almost entirely from liquid air.

Discovery and Naming

What is air? Ancient peoples thought deeply about that question. And that should not be surprising. It is easy to see how essential air is to many processes. Objects cannot burn without air. Human life cannot survive without air. In fact, ancient peoples thought air must be an "element." But they used the word "element" differently than modern scientists do. To ancient people, an element was something that was very important and basic. Air fit that description, along with fire, water, and earth.

They often thought of air as an element in the modern sense—that it was as simple a material as could be found. Yet, some early scholars believed otherwise. For example, some Chinese scholars, as early as the eighth century CE, thought of air as having two parts. They called these parts the *yin* and *yang* of air. The properties of the Chinese yin and yang can be compared to the properties of oxygen and **nitrogen**.

The first person in Western Europe to describe the "parts" of air was Italian artist and scientist Leonardo da Vinci (1452–1519). Leonardo pointed out that air is not entirely used up when something is burned in it. He said that air must consist, therefore, of two parts: one part that is consumed in burning and one part that is not.

For many years, Leonardo's ideas were not very popular among scholars. One problem was that early chemists did not have very good equipment. It was difficult for them to collect samples of air and then to study it.

In the early 1700s, chemists began to find out more about air, but in a somewhat roundabout way. For example, in 1771 and 1772, Scheele studied the effect of heat on a number of different compounds. In one experiment, he used silver carbonate (Ag_2CO_3), mercury carbonate ($HgCO_3$), and magnesium nitrate (Mg $(NO_3)_2$). When he heated these compounds,

Joseph Priestley.

he found that a gas was produced. He then studied the properties of that gas. He found that flames burned brightly in the gas. He also found that animals could live when placed in the gas. Without knowing it, Scheele had discovered oxygen.

About two years later, Joseph Priestley conducted similar experiments by heating mercury(II) oxide (HgO) in a flame. The compound broke down, producing liquid mercury metal and a gas:

$$2HgO \text{ —heat} \rightarrow 2Hg + O_2$$

When Priestley tested the new gas, he found the same properties that Scheele had described.

Priestley even tried breathing the new gas he had produced. His description of that experience has now become famous:

> The feeling of it [the new gas, oxygen] to my lungs was not sensibly different from that of common air, but I fancied that my breast felt peculiarly light and easy for some time afterwards. Who can tell but that, in time, this pure air may become a fashionable article in luxury? Hitherto only two mice and myself have had the privilege of breathing it.

Some people think Scheele should get credit for discovering oxygen. He completed his experiments earlier than did Priestley. But his publisher was very slow in printing Scheele's reports. They actually came out after

Antoine-Laurent Lavoisier.

Priestley's reports. So most historians agree that Scheele and Priestley should share credit for discovering oxygen.

Neither Scheele nor Priestley fully understood the importance of their discovery. That step was taken by French chemist Antoine-Laurent Lavoisier (1743–1794). Lavoisier was the first person to declare that the new gas was an element. He was also the first person to explain how oxygen is involved in burning.

In addition, Lavoisier suggested a name for the gas. That name, oxygen, comes from Greek words that mean "acidic" *(oxy-)* and "forming" *(-gen)*. Lavoisier chose the name because he thought that all acids contain oxygen. Therefore, the new element was responsible for "forming acids." In this one respect, however, Lavoisier was wrong. All acids do not contain oxygen, although some do.

Physical Properties

Oxygen is a colorless, odorless, tasteless gas. It changes from a gas to a liquid at a temperature of –297.33°F (–182.96°C). The liquid formed has a slightly bluish color to it. Liquid oxygen can then be solidified or frozen at a temperature of –361.2°F (–218.4°C). The density of oxygen is 1.429 grams per liter. By comparison, the density of air is about 1.29 grams per liter.

Oxygen exists in three allotropic forms. Allotropes are forms of an element with different physical and chemical properties. The three allotropes of oxygen are normal oxygen, or diatomic oxygen, or dioxygen; nascent, atomic, or monatomic oxygen; and ozone, or triatomic oxygen. The three allotropes differ from each other in a number of ways.

First, they differ on the simplest level of atoms and molecules. The oxygen that we are most familiar with in the atmosphere has two atoms in every molecule. Chemists show this by writing the formula as O_2. The small "2" means "two atoms per molecule."

By comparison, nascent oxygen has only one atom per molecule. The formula is simply O, or sometimes (O). The parentheses indicate that

The Father of Modern Chemistry

Antoine-Laurent Lavoisier is often called the father of modern chemistry. He was given that title for a number of reasons. The most important reason is the explanation he discovered for the process of combustion (burning).

Prior to Lavoisier's research, chemists thought that a burning object gave off a substance to the air. They called that substance phlogiston. When wood burned, for example, chemists said that phlogiston escaped from the wood to the air.

Lavoisier showed that this idea was incorrect. When something burns, it actually combines with oxygen in the air. Combustion, Lavoisier said, is really just oxidation (the process by which something combines with oxygen).

This discovery gave chemists a whole new way to look at chemical changes. The phlogiston theory gradually began to die out. Many of the ideas used in modern chemistry began to develop.

Lavoisier led an unusually interesting life. He was an avid chemist who carried out many experiments. But he also had a regular job as a tax collector. His job was to visit homes and businesses to collect taxes. This did not make him a very popular man.

Lavoisier also made some important enemies early in his life. One of these enemies was Jean-Paul Marat (1743–1793). Marat thought of himself as a scientist and applied for membership in the French Academy of Scientists. Lavoisier voted against Marat's application. He said that Marat's research was not very good.

Less than a decade later, Lavoisier had reason to regret that decision. Marat had become a leader in the French Revolution (1774–1815). He accused Lavoisier of plotting against the revolution. He also said that Lavoisier was carrying out dangerous secret experiments.

These accusations were not true. But Marat was now a very powerful man. He was able to have Lavoisier convicted of the charges against him. On May 8, 1794, Lavoisier was beheaded and buried in an unmarked grave. Some people have said that Lavoisier's death was the worst single consequence of the French Revolution.

nascent oxygen does not exist very long under normal conditions. It has a tendency to form dioxygen:

$$O + O \rightarrow O_2$$

That is, dioxygen is the normal condition of oxygen at room temperature.

The third allotrope of oxygen, ozone, has three atoms in each molecule. The chemical formula is O_3. Like nascent oxygen, ozone does not exist for very long under normal conditions. It tends to break down and form dioxygen:

$$2O_3 \rightarrow 3O_2$$

Ozone does occur in fairly large amounts under special conditions. For example, there is an unusually large amount of ozone in Earth's upper atmosphere. That ozone layer is important to life on Earth. It shields out harmful radiation that comes from the sun. Ozone is also sometimes found closer to Earth's surface. It is produced when gasoline is burned in cars and trucks. It is part of the condition known as air pollution. Ozone at ground level is not helpful to life, and may cause health problems for plants, humans, and other animals.

The physical properties of ozone are somewhat different from those of dioxygen. It has a slightly bluish color as both a gas and a liquid. It changes to a liquid at a temperature of –169.4°F (–111.9°C) and from a liquid to a solid at –315°F (–193°C). The density is 2.144 grams per liter.

Chemical Properties

Oxygen's most important chemical property is that it supports combustion. That is, it helps other objects to burn. The combustion (burning) of charcoal is an example. Charcoal is nearly pure **carbon** (C):

$$C + O_2 \rightarrow CO_2$$

Oxygen also combines with elements at room temperature. Rusting is an example. Rusting is a process by which a metal combines with oxygen. When iron rusts, it combines with oxygen:

$$4Fe + 3O_2 \rightarrow 2Fe_2O_3$$

Oxygen also reacts with many compounds. Decay is an example. Decay is the process by which once-living material combines with oxygen. The products of decay are mainly carbon dioxide (CO_2) and water (H_2O):

$$\text{dead matter} + O_2 \rightarrow CO_2 + H_2O$$

(The chemical formula for "dead matter" is too complicated to use here.)

Oxygen itself does not burn. A lighted match in a container of pure oxygen burns much brighter, but the oxygen does not catch fire.

Occurrence in Nature

Oxygen occurs mainly as an element in the atmosphere. It makes up 20.948 percent of the atmosphere. It also occurs in oceans, lakes, rivers, and ice caps in the form of water. Nearly 89 percent of the weight of water is oxygen. Oxygen is also the most abundant element in Earth's crust. Its abundance is estimated at about 45 percent in the earth. That makes it almost twice as abundant as the next most common element, **silicon**.

Oxygen occurs in all kinds of minerals. Some common examples include the oxides, carbonates, nitrates, sulfates, and phosphates. Oxides are chemical compounds that contain oxygen and one other element. Calcium oxide, or lime or quicklime (CaO), is an example. Carbonates are compounds that contain oxygen, carbon, and at least one other element. Sodium carbonate, or soda, soda ash, or sal soda (Na_2CO_3), is an example. It is often found in detergents and cleaning products.

Nitrates, sulfates, and phosphates also contain oxygen and other elements. The other elements in these compounds are nitrogen, sulfur, or **phosphorus** plus one other element. Examples of these compounds are **potassium** nitrate, or saltpeter (KNO_3); magnesium sulfate, or Epsom salts ($MgSO_4$); and calcium phosphate ($Ca_3(PO_4)_2$).

Isotopes

There are three naturally occurring isotopes of oxygen: oxygen-16, oxygen-17, and oxygen-18. Isotopes are two or more forms of an element. Isotopes differ from each other according to their mass number. The number written to the right of the element's name is the mass number. The mass number represents the number of protons plus neutrons in the nucleus of an atom of the element. The number of protons determines the element, but the number of neutrons in the atom of any one element can vary. Each variation is an isotope.

Ten radioactive isotopes of oxygen are known also. A radioactive isotope is one that breaks apart and gives off some form of radiation. Radioactive isotopes are produced when very small particles are fired at atoms. These particles stick in the atoms and make them radioactive.

None of the radioactive isotopes of oxygen has any commercial use.

Extraction

Oxygen is made from liquid air. Liquid air is made by cooling normal atmospheric air to very low temperatures. As the temperature drops, the gases contained in air turn into liquids. At –297.33°F (–182.96°C), oxygen changes from a gas into a liquid. At –320.42°F (–195.79°C), nitrogen changes from a gas into a liquid. And so on. Eventually, all the gases in air can be made to liquefy (change into a liquid).

But the reverse process also takes place. Suppose liquid air in a container warms up slowly. When its temperature reaches –320.42°F (–195.79°C), liquid nitrogen changes back to a gas. A container can be put into place to catch the nitrogen as it boils off the liquid air. When the temperature reaches –297.33°F (–182.96°C), oxygen changes from a liquid back to a gas. Another container can be put into place. The escaping oxygen can be collected. Oxygen with a purity of 99.995 percent can be made by this method. It is the only method by which oxygen is made for commercial purposes.

Uses

Many people are familiar with oxygen to help preserve lives. In some cases, people are not able to breathe on their own. Conditions such as emphysema damage the lungs. Oxygen cannot pass through the lungs into the bloodstream. One way to treat this condition is to force oxygen into the lungs with a pump.

The same method is used to treat other medical conditions. For example, carbon monoxide poisoning occurs when carbon monoxide gas

Standard firefighting equipment includes compressed air tanks to help firefighters breathe in smoke-filled situations. IMAGE COPYRIGHT 2009, M. NORRIS. USED UNDER LICENSE FROM SHUTTERSTOCK.COM.

gets into the bloodstream. Auto exhaust, poorly maintained oil furnaces, and wood fires produce carbon monoxide. The carbon monoxide replaces oxygen in the blood. Cells get carbon monoxide instead of oxygen. But they cannot use carbon monoxide, so they begin to die. Forcing oxygen into the blood can reverse some of the damage. In high enough amounts, it can force the carbon monoxide out of the blood and cells can recover.

Oxygen has other interesting uses. For example, it is used in rocket fuels. It is combined with hydrogen in the rocket engines. When hydrogen and oxygen combine, they give off very large amounts of energy. The energy is used to lift the rocket into space.

Metal production accounts for the greatest percentage of oxygen use. For example, oxygen is used to burn off carbon and other impurities that are in iron to make steel. A small amount of these impurities may be desirable in steel, but too much makes it brittle and unusable. The carbon and other impurities are burned off in steel-making by blasting oxygen through molten iron.

The unmanned Saturn rocket, shortly after takeoff on January 22, 1968. The combination of oxygen and hydrogen creates enough energy to lift the rocket into space. NATIONAL AERONAUTICS AND SPACE ADMINISTRATION (NASA).

Two chemical changes that take place during steel-making are shown below:

$$2C + O_2 \rightarrow CO_2$$

$$Si + O_2 \rightarrow SiO_2$$

The carbon dioxide escapes from the steel-making furnace as a gas. The silicon dioxide (SiO_2) forms slag. Slag is a crusty, metallic material that is scraped off after the steel is produced. Other impurities removed by a blast of oxygen are sulfur, phosphorus, **manganese**, and other metals.

Oxygen is also used in the production of such metals as **copper**, **lead**, and **zinc**. These metals occur in the earth in the form of sulfides, such as copper sulfide (CuS), lead sulfide (PbS), and zinc sulfide (ZnS). The first step in recovering these metals is to convert them to oxides:

$$2CuS + 3O_2 \rightarrow 2CuO + 2SO_2$$

$$2PbS + 3O_2 \rightarrow 2PbO + 2SO_2$$

$$2ZnS + 3O_2 \rightarrow 2ZnO + 2SO_2$$

The oxides are then heated with carbon to make the pure metals:

$$2CuO + C \rightarrow 2Cu + CO_2$$

$$2PbO + C \rightarrow 2Pb + CO_2$$

$$2ZnO + C \rightarrow 2Zn + CO_2$$

Another use of oxygen is in high-temperature torches. The oxyacetylene torch, for example, produces heat by burning acetylene gas (C_2H_2) in pure oxygen. The torch can produce temperatures of 5,400°F (3,000°C) and cut through steel and other tough alloys.

Oxygen is also used in the chemical industry as a beginning material in making some very important compounds. Sometimes, the steps to get from oxygen to the final compound are lengthy. As an example, ethylene gas (C_2H_4) can be treated with oxygen to form ethylene oxide (CH_2CH_2O):

$$2CH_2H_4 + O_2 \rightarrow 2CH_2H_4O$$

About 60 percent of ethylene oxide produced is made into ethylene glycol ($CH_2CH_2(OH)_2$). Ethylene glycol is used in antifreeze and as a starting point in making polyester fibers, film, plastic containers, bags, and packaging materials.

Compounds

Thousands of oxygen compounds have important commercial uses. Many of these compounds are discussed under other elements.

Health Effects

Nearly all organisms require oxygen—bacteria, plants, and animals. Humans, for example, can go weeks without food. They can survive for many days without water. But they cannot survive more than a few minutes without oxygen.

Oxygen is used by the cells of animal bodies. It is used to oxidize chemicals and produce energy that cells need to stay alive. Without oxygen, cells begin to die in minutes.

Where to Learn More

The following list focuses on works written for readers of middle school and high school age. Books aimed at adult readers have been included when they are especially important in providing information or analysis that would otherwise be unavailable, or because they have become classics.

Books

Atkins, P. W. *The Periodic Kingdom: A Journey into the Land of the Chemical Elements.* New York: HarperCollins, 1997.

Ball, Philip. *The Ingredients: A Guided Tour of the Elements.* New York: Oxford University Press, 2003.

Emsley, John. *Nature's Building Blocks: An A-Z Guide to the Elements.* New York: Oxford University Press, 2003.

Gray, Theodore W., and Nick Mann. *The Elements: A Visual Exploration of Every Known Atom in the Universe.* New York: Black Dog & Leventhal Publishers, 2009.

Greenwood, N. N., and A. Earnshaw. *Chemistry of the Elements, 2nd Edition.* Oxford: Pergamon Press, 2006.

Harvey, David I. *Deadly Sunshine: The History and Fatal Legacy of Radium.* Stroud, Gloucestershire, UK: Tempus, 2005.

Krebs, Robert E. *The History and Use of Our Earth's Chemical Elements: A Reference Guide, 2nd Edition.* Westport, CT: Greenwood Publishing Group, 2006.

Lew, Kristi. *Radium.* New York: Rosen Central, 2009.

Lewis, Richard J., Sr. *Hawley's Condensed Chemical Dictionary, 15th edition.* New York: Wiley Interscience, 2007.

Miller, Ron. *The Elements.* Minneapolis: Twentieth Century Books, 2006.

Minerals Information Team. *Minerals Yearbook.* Reston, VA: U.S. Geological Survey, 2009.

Nachaef, N. *The Chemical Elements: The Exciting Story of Their Discovery and of the Great Scientists Who Found Them.* Jersey City, NJ: Parkwest Publications, 1997.

Ruben, Samuel. *Handbook of the Elements, 3rd Edition.* La Salle, IL: Open Court Publishing Company, 1999.

Stwertka, Albert. *A Guide to the Elements, 2nd Edition.* New York: Oxford University Press, 2002.

Weeks, Mary Elvira. *Discovery of the Elements.* Whitefish, MT: Kessinger Publishing Reprints, 2003.

Periodicals

Barber, R. C., et al. "Discovery of the Element with Atomic Number 112." *Pure and Applied Chemistry* (August 13, 2009): 1,331–1,343.

Barceloux, D. G. "Chromium." *Journal of Toxicology. Clinical Toxicology* 37 (1999): 173–194.

Biever, Celeste. "The Tick, Tock of a Strontium Clock Gets Steadier." *New Scientist* (December 9, 2006): 32.

Camparo, James. "The Rubidium Atomic Clock and Basic Research." *Physics Today* (November 2007): 33–39.

Cotton, S. A. "The Actinides." *Annual Reports Section A (Inorganic Chemistry)* 105 (2009): 287–296.

Cotton, S. A. "Titanium, Zirconium, and Hafnium." *Annual Reports Section A (Inorganic Chemistry)* 105 (2009): 177–187.

Crans, Debbie D., Jason J. Smee, Ernestas Gaidamauskas, and Luqin Yang. "The Chemistry and Biochemistry of Vanadium and the Biological Activities Exerted by Vanadium Compounds." *Chemical Reviews* (January 2004): 849–902.

Croswell, Ken. "Fluorine: An Element-ary Mystery." *Sky and Telescope* (September 2003): 30–37.

Fricker, Simon P. "Therapeutic Applications of Lanthanides." *Chemical Society Reviews* 35 (2006): 524–533.

Garelick, Hemda, Huw Jones, Agnieszka Dybowska, and Eugenia Valsami-Jones. "Arsenic Pollution Sources." *Reviews of Environmental Contamination and Toxicology* 197 (2008): 17–60.

Geim, Andre K., and Philip Kim. "Carbon Wonderland." *Scientific American* (April 2008): 90–97.

Gomez, Gerda E. "Lithium Treatment: Present and Future." *Journal of Psychosocial Nursing and Mental Health Services* (August 2001): 31–37.

Gruber, Nicolas, and James N. Galloway. "An Earth-system Perspective of the Global Nitrogen Cycle." *Nature* (January 17, 2008): 293–296.

Guerrera, Mary P., Stella Lucia Volpe, and James Mao Jun. "Therapeutic Uses of Magnesium." *American Family Physician* (July 15, 2009): 157–162.

Kolanz, Marc. "Beryllium History and Public Policy." *Public Health Reports* (July 2008): 423–428.

Ljung, Karin, and Marie Vahter. "Time to Re-Evaluate the Guideline Value for Manganese in Drinking Water?" *Environmental Health Perspectives* (November 2007): 1,533–1,538.

Martin, Richard. "Uranium Is So Last Century—Enter Thorium, the New Green Nuke." *WIRED* (January 2010).

Nielsen, Forrest H. "Is Boron Nutritionally Relevant?" *Nutrition Reviews* (April 2008): 183–191.

Ogden, Joan. "High Hopes for Hydrogen." *Scientific American* (September 2006): 94–101.

Parr, J. "Carbon, Silicon, Germanium, Tin and Lead." *Annual Reports Section A (Inorganic Chemistry)* 105 (2009): 117–139.

Roney, Nicolette, et al. "ATSDR Evaluation of Potential for Human Exposure to Zinc." *Toxicology and Industrial Health* 22 (2006): 423–493.

Satyapal, Sunita, John Petrovic, and George Thomas. "Gassing Up with Hydrogen." *Scientific American* (April 2007): 80–87.

Stein, Andreas. "Materials Science: Germanium Takes Holey Orders." *Nature* (June 29, 2006): 1,055–1,056.

Utiger, Robert D. "Iodine Nutrition—More Is Better." *New England Journal of Medicine* (June 29, 2006): 2,819–2,821.

Vaccari, David A. "Phosphorus: A Looming Crisis." *Scientific American* (June 2009): 54–59.

van Delft, Dirk. "Little Cup of Helium, Big Science." *Physics Today* (March 2008): 36–42.

Web Sites

Air Chek, Inc. "The Radon Information Center." http://www.radon.com/ (accessed February 9, 2010).

American Chemistry Council, Chlorine Chemistry Division. http://www.americanchemistry.com/chlorine/ (accessed February 9, 2010).

American Elements. "Sodium." http://www.americanelements.com/na.html (accessed February 9, 2010).

American Elements. "Zinc." http://www.americanelements.com/zn.html (accessed February 9, 2010).

BBC News. "New Element Named 'Copernicium.'" http://news.bbc.co.uk/2/hi/science/nature/8153596.stm (accessed February 9, 2010).

Borland, John. "Scientists Use Superconducting Cyclotron to Make Super-Heavy Metals" (October 24, 2007). http://www.wired.com/science/discoveries/news/2007/10/heavymetalisotopes (accessed February 9, 2010).

Canadian Nuclear Association. "How Is Nuclear Technology Used in Smoke Detectors?" http://www.cna.ca/english/pdf/NuclearFacts/18-NuclearFacts-smokedetectors.pdf (accessed February 9, 2010).

EnvironmentalChemistry.com. "Periodic Table of Elements." http://environmentalchemistry.com/yogi/periodic/ (accessed February 9, 2010).

Excalibur Mineral Company. "Tin Mineral Data." http://webmineral.com/data/Tin.shtml (accessed February 9, 2010).

GreenFacts. "Scientific Facts on Fluoride." http://www.greenfacts.org/en/fluoride/index.htm (accessed February 9, 2010).

Holden, Norman E. "History of the Origin of the Chemical Elements and Their Discoverers." http://www.nndc.bnl.gov/content/elements.html#boron (accessed February 9, 2010).

Indium Metal Information Center. http://www.indium.com/products/indiummetal.php (accessed February 9, 2010).

International Atomic Energy Agency. "Number of Reactors in Operation Worldwide." http://www.iaea.org/cgi-bin/db.page.pl/pris.oprconst.htm (accessed February 9, 2010).

International Platinum Group Metals Association. "Iridium." http://www.ipa-news.com/pgm/iridium/index.htm (accessed February 9, 2010).

International Union of Pure and Applied Chemistry (IUPAC). "The Chemical Elements." http://old.iupac.org/general/FAQs/elements.html (accessed February 9, 2010).

Lenntech. "Periodic Table." http://www.lenntech.com/periodic/periodic-chart.htm (accessed February 9, 2010).

Los Alamos National Laboratory. Chemical Division. "Periodic Table of the Elements." http://periodic.lanl.gov/default.htm (accessed February 9, 2010).

Mineral Information Institute. "Titanium." http://www.mii.org/Minerals/phototitan.html (accessed February 9, 2010).

Mineral Information Institute. "Vanadium." http://www.mii.org/Minerals/photovan.html (accessed February 9, 2010).

Nuclear Regulatory Commission. "Map of the United States Showing Locations of Operating Nuclear Power Reactors." http://www.nrc.gov/info-finder/reactor/ (accessed February 9, 2010).

ResponsibleGold.org. http://www.responsiblegold.org/ (accessed February 9, 2010).

Royal Society of Chemistry. "Radium—Ra." http://www.rsc.org/chemsoc/visualelements/Pages/data/radium_data.html (accessed February 9, 2010).

Royal Society of Chemistry. "Visual Interpretation of the Table of Elements." http://www.rsc.org/chemsoc/visualelements/index.htm (accessed February 9, 2010).

Silver Institute. "Silver Facts: Silver as an Element." http://www.silverinstitute.org/silver_element.php (accessed February 9, 2010).

Smart Elements.com. "Elements." http://www.smart-elements.com/?arg=pse&lid=11&PHPSESSID=3760317e40a3f4cf31f2a87bc17cc413 (accessed February 9, 2010).

U.S. Department of Energy, Lawrence Berkeley National Laboratory. "Superheavy Element 114 Confirmed: A Stepping Stone to the Island of Stability." http://newscenter.lbl.gov/press-releases/2009/09/24/114-confirmed/ (February 9, 2010).

U.S. Department of Health and Human Services. "Dietary Guidelines for Americans." http://www.health.gov/dietaryguidelines/ (accessed February 9, 2010).

U.S. Environmental Protection Agency. "Carbon Dioxide." http://epa.gov/climatechange/emissions/co2.html (accessed February 9, 2010).

U.S. Environmental Protection Agency. "A Citizen's Guide to Radon: The Guide to Protecting Yourself and Your Family from Radon" (January 2009). http://www.epa.gov/radon/pdfs/citizensguide.pdf (accessed February 9, 2010).

U.S. Environmental Protection Agency. "An Introduction to Indoor Air Quality: Carbon Monoxide." http://www.epa.gov/iaq/co.html (accessed February 9, 2010).

U.S. Geological Survey Minerals Yearbook. "Bismuth Statistics and Information." http://minerals.usgs.gov/minerals/pubs/commodity/bismuth/ (accessed February 9, 2010).

U.S. Geological Survey Minerals Yearbook. "Lithium Statistics and Information." http://minerals.usgs.gov/minerals/pubs/commodity/lithium/ (accessed February 9, 2010).

U.S. Geological Survey Minerals Yearbook. "Nickel Statistics and Information." http://minerals.usgs.gov/minerals/pubs/commodity/nickel/ (accessed February 9, 2010).

U.S. Geological Survey Minerals Yearbook. "Nitrogen Statistics and Information." http://minerals.usgs.gov/minerals/pubs/commodity/nitrogen/ (accessed February 9, 2010).

U.S. Geological Survey Minerals Yearbook. "Platinum-Metals Statistics and Information." http://minerals.usgs.gov/minerals/pubs/commodity/platinum/ (accessed February 9, 2010).

U.S. Geological Survey Minerals Yearbook. "Tin Statistics and Information." http://minerals.usgs.gov/minerals/pubs/commodity/tin/ (accessed February 9, 2010).

U.S. Geological Survey Minerals Yearbook. "Vanadium Statistics and Information." http://minerals.usgs.gov/minerals/pubs/commodity/vanadium/ (accessed February 9, 2010).

University of Nottingham. "The Periodic Table of Videos." http://www.periodicvideos.com (accessed February 9, 2010).

WebQC.com. "Periodic Table of Chemical Elements." http://www.webqc.org/periodictable.php (accessed February 9, 2010).

World Information Service on Energy. "Uranium Project." http://www.wise-uranium.org/ (accessed February 9, 2010).

Organizations

Aluminum Association, 1525 Wilson Blvd., Suite 600, Arlington, VA 22209; Phone: 703-358-2960. Internet: www.aluminum.org.

American Chemistry Council, 1300 Wilson Blvd., Arlington, VA 22209. Phone: 703-741-5000. Internet: http://www.americanchemistry.com/.

American Hydrogen Association, 2350 W. Shangri La, Phoenix, AZ 85029. Phone: 602-328-4238. Internet: http://www.clean-air.org/.

American Iron and Steel Institute, 1140 Connecticut Ave., NW, Suite 705, Washington, DC 20036. Phone: 202-452-7100. Internet: http://www.steel.org.

Australian Uranium Association, GPO Box 1649, Melbourne, VIC 3001, Australia. Phone: +61 3 8616 0440. Internet: http://aua.org.au/.

Copper Development Association, 260 Madison Ave., New York, NY 10016. Phone: 212-251-7200. Internet: http://www.copper.org/.

International Atomic Energy Agency (IAEA), Vienna International Centre, PO Box 100, 1400 Vienna, Austria. Phone: (+431) 2600-0. Internet: http://www.iaea.org/.

International Lead Association. 17a Welbeck Way, London, United Kingdom, W1G 9YJ. Phone: +44 (0)207 499 8422. Internet: http://www.ila-lead.org/.

International Magnesium Association, 1000 N. Rand Rd., Suite 214, Wauconda, IL 60084. Phone: 847-526-2010. Internet: http://www.intlmag.org/.

International Molybdenum Association, 4 Heathfield Terrace, London, United Kingdom, W4 4JE. Phone: +44 (0)207 871 1580. Internet: http://www.imoa.info/.

International Platinum Group Metals Association, Schiess-Staett-Strasse 30, 80339 Munich, Germany. Phone: (+49) (0)89 519967-70. Internet: http://www.ipa-news.com/.

International Union of Pure and Applied Chemistry, IUPAC Secretariat, PO Box 13757, Research Triangle Park, NC 27709-3757. Internet: www.iupac.org/.

International Zinc Association—America, 2025 M Street, NW, Suite 800, Washington DC 20036. Phone: 202-367-1151. Internet: http://www.zinc.org/.

Los Alamos National Laboratory, PO Box 1663, Los Alamos, NM 87545. Phone: 505-667-5061. Internet: www.lanl.gov/.

Mineral Information Institute, 8307 Shaffer Pkwy., Littleton, CO 80127. Phone: 303-277-9190. Internet: http://www.mii.org.

National Hydrogen Association, 1211 Connecticut Ave., NW, Suite 600, Washington, DC 20036-2701. Phone: 202-223-5547. Internet: http://www.hydrogenassociation.org/.

Nickel Institute, Brookfield Place, 161 Bay St., Suite 2700, Toronto, ON, Canada M5J 2S1. Phone: 416-591-7999. Internet: http://www.nidi.org/.

Royal Society of Chemistry, Burlington House, Piccadilly, London, United Kingdom W1J 0BA. Phone: +44 (0)207 437 8656. Internet: http://www.rsc.org/.

Selenium-Tellurium Development Association, Cavite Economic Zone, Lot 6, Blk 1, Phase II Rosario, Cavite, Philippines. Phone: +63 46 437 2526. Internet: http://www.stda.org/.

Sulphur Institute, 1140 Connecticut Ave., NW, Suite 612, Washington, DC 20036. Phone: 202-331-9660. Internet: http://www.sulphurinstitute.org/.

Tantalum-Niobium International Study Center, Chaussée de Louvain 490, 1380 Lasne, Belgium. Phone: +32 2 649 51 58. Internet: http://tanb.org/.

U.S. Department of Energy (DOE), Lawrence Berkeley National Laboratory, 1 Cyclotron Rd., Berkeley, CA 94720. Phone: 510-486-4000. Internet: www.lbl.gov/.

U.S. Environmental Protection Agency (EPA), Ariel Rios Building, 1200 Pennsylvania Ave., NW, Washington, DC 20460. Phone: 202-272-0167. Internet: www.epa.gov/.

U.S. Geological Survey (USGS) National Center, 12201 Sunrise Valley Dr., Reston, VA 20192. Phone: 703-648-5953. Internet: www.usgs.gov/.

Index

Italic type indicates volume numbers;
boldface indicates main entries.
Illustrations are marked by (ill.)

Bromine-79, *1*: 76

Bromine-81, *1*: 76

Bronze, *1*: 149, 150, 154

Bronze Age, *1*: 150

Brookhaven National Laboratory, *1*: 40–41

Brunetti, Rita, *3*: 468

Buckminsterfullerene. *See* Fullerene

Buckyballs. *See* Fullerene

Bunsen, Robert, *1*: 119, 120, 121 (ill.); *2*: 262; *3*: 501, 502

Bureau of Mines, U.S., *2*: 244

C

C. *See* Carbon

Ca. *See* Calcium

Cade, John, *2*: 319

Cadmium, *1*: **79–84**, 79 (ill.), 81 (ill.), 83 (ill.)

 batteries (nicad or Ni-Cd), *1*: 83, 83 (ill.)

 discovery and naming of, *1*: 80

 health effects of, *1*: 84

 isotopes of, *1*: 82

 uses of, *1*: 83

Cadmium-106, *1*: 82

Cadmium-108, *1*: 82

Cadmium-109, *1*: 82

Cadmium-110, *1*: 82

Cadmium-111, *1*: 82

Cadmium-112, *1*: 82

Cadmium-114, *1*: 82

Cadmium-116, *1*: 82

Cadmium oxide, *1*: 81

Cadmium selenide, *1*: 83–84

Cadmium sulfide, *1*: 82, 84–85

Caesium. *See* Cesium

Calamine lotion. *See* Zinc oxide

Calavarite, *2*: 228

Calcite, *1*: 43, 88

Calcium, *1*: **85–93**, 85 (ill.)

 argon and, *1*: 28

 barium and, *1*: 43

 bismuth and, *1*: 62

 carbon and, *1*: 105

 cesium and, *1*: 123

 compounds of, *1*: 44, 85–87, 88, 90–93

discovery and naming of, *1*: 43, 85–88

dysprosium and, *1*: 168

extraction of, *1*: 90

fluorine and, *1*: 192

health effects of, *1*: 92, 93

isotopes of, *1*: 89–90

natural occurrence of, *1*: 88, 89, 89 (ill.), 91 (ill.)

neodymium and, *2*: 361

in periodic table, *1*: 53

strontium and, *3*: 558

uses of, *1*: 85–86, 89–93

Calcium-40, *1*: 89

Calcium-42, *1*: 89

Calcium-43, *1*: 89

Calcium-44, *1*: 89

Calcium-45, *1*: 90

Calcium-46, *1*: 89

Calcium-47, *1*: 90

Calcium-48, *1*: 89

Calcium alginate, *1*: 92

Calcium arsenate, *1*: 92

Calcium borate. *See* Colemanite

Calcium carbide, *1*: 92

Calcium carbonate. *See* Limestone

Calcium chloride, *1*: 85, 90, 92

Calcium cyclamate, *1*: 92

Calcium fluoride, *1*: 191, 193

Calcium gluconate, *1*: 92

Calcium hydroxide, *1*: 92

Calcium hypochlorite, *1*: 92

Calcium magnesium carbonate, *1*: 43; *2*: 326, 328

Calcium oxide, *1*: 88, 91, 92

Calcium permanganate, *1*: 92

Calcium phosphate, *1*: 92

Calcium phosphide, *1*: 93

Calcium silicate, *1*: 91; *2*: 375, 413

Calcium stearate, *1*: 93

Calcium sulfate. *See* Gypsum

Calcium sulfite, *1*: 92

Calcium tungstate, *1*: 93

California

 californium and, *1*: 95

 cobalt in, *1*: 144

 gold in, *2*: 226–227, 226 (ill.), 229

 hydrogen in, *2*: 259

 mercury in, *2*: 344

 selenium in, *3*: 528–529

E

Ears, *2*: 207

Earths, *2*: 205, 248

Egypt, elements in, *2*: 224, 225 (ill.), 284, 307

Ehrlich, Paul, *1*: 35

Einstein, Albert, *1*: 172, 172 (ill.)

Einsteinium, *1*: **171–173**, 171 (ill.)
   discovery and naming of, *1*: 171–172
   isotopes of, *1*: 172–173

Einsteinium-254, *1*: 172, 173

Ekaboron. *See* Scandium

Ekeberg, Anders Gustaf, *3*: 569, 570

El Dorado, *2*: 226–227

Electricity
   aluminum and, *1*: 7, 8
   americium and, *1*: 17
   argon and, *1*: 28
   beryllium and, *1*: 56–57
   cesium and, *1*: 122, 123
   chromium and, *1*: 140
   cobalt and, *1*: 145
   copper and, *1*: 149, 150, 153–154
   curium and, *1*: 163
   Davy, Humphry and, *1*: 87
   gallium and, *2*: 213–215
   germanium and, *2*: 219
   gold and, *2*: 227
   helium and, *2*: 245
   hydrogen and, *2*: 256
   indium and, *2*: 265
   iron and, *2*: 283
   krypton and, *2*: 297–298
   mercury and, *2*: 343–344
   phosphorus and, *3*: 424
   polonium and, *3*: 447
   rhodium and, *3*: 498
   selenium and, *3*: 524
   silver and, *3*: 543–544

Electrochemistry, *1*: 87

Electrodes, *1*: 87

Electrolysis, *1*: 87, 153; *2*: 310; *3*: 451, 452–453, 543

Electronics and electrical equipment
   aluminum in, *1*: 5, 8, 11

   arsenic in, *1*: 34
   beryllium in, *1*: 56
   boron in, *1*: 70
   cadmium in, *1*: 83
   copper in, *1*: 153–154
   dysprosium in, *1*: 168, 169
   gallium in, *2*: 210, 213–215
   germanium in, *2*: 217
   gold in, *2*: 230
   indium in, *2*: 261, 264, 265 (ill.)
   iridium in, *2*: 281
   lead in, *2*: 311
   lithium in, *2*: 318–319
   manganese in, *2*: 338–339
   mercury in, *2*: 346
   neodymium in, *2*: 361, 361 (ill.)
   nickel in, *2*: 379, 380–381, 381 (ill.)
   rubidium in, *3*: 504
   silicon in, *3*: 531–532, 534, 535
   silver in, *3*: 544
   strontium in, *3*: 555
   tantalum in, *3*: 572
   See also specific devices

Electroplating. *See* Coatings

Emanium. *See* Actinium

Emantion. *See* Radon

Emeralds, *1*: 54, 57 (ill.)

Emissions. *See* Pollution

Energy conservation, *2*: 258–259, 265–266

Engines, *1*: 117, 141; *2*: 312

Enrich bulb, *2*: 361

Environmental effects
   of arsenic, *1*: 34, 35
   of bromine, *1*: 77–78
   of cadmium, *1*: 79, 83, 84
   of carbon, *1*: 110, 111–112, 112 (ill.)
   of cesium, *1*: 122 (ill.)
   of chlorine, *1*: 130, 132
   of lead, *2*: 310–311, 312
   of mercury, *2*: 345, 346
   of nitrogen, *2*: 399
   of phosphorus, *3*: 427–428
   radon and, *3*: 485
   selenium and, *3*: 528–530
   of strontium, *3*: 558

F

I

P

U